T0292576

Studies in Big Data

Volume 17

Series editor

Janusz Kacprzyk, Polish Academy of Sciences, Warsaw, Poland
e-mail: kacprzyk@ibspan.waw.pl

About this Series

The series "Studies in Big Data" (SBD) publishes new developments and advances in the various areas of Big Data- quickly and with a high quality. The intent is to cover the theory, research, development, and applications of Big Data, as embedded in the fields of engineering, computer science, physics, economics and life sciences. The books of the series refer to the analysis and understanding of large, complex, and/or distributed data sets generated from recent digital sources coming from sensors or other physical instruments as well as simulations, crowd sourcing, social networks or other internet transactions, such as emails or video click streams and other. The series contains monographs, lecture notes and edited volumes in Big Data spanning the areas of computational intelligence incl. neural networks, evolutionary computation, soft computing, fuzzy systems, as well as artificial intelligence, data mining, modern statistics and Operations research, as well as self-organizing systems. Of particular value to both the contributors and the readership are the short publication timeframe and the world-wide distribution, which enable both wide and rapid dissemination of research output.

More information about this series at http://www.springer.com/series/11970

Bhabani Shankar Prasad Mishra
Satchidananda Dehuri · Euiwhan Kim
Gi-Name Wang
Editors

Techniques
and Environments
for Big Data Analysis

Parallel, Cloud, and Grid Computing

 Springer

Editors
Bhabani Shankar Prasad Mishra
School of Computer Engineering
KIIT University
Bhubaneswar, Odisha
India

Euiwhan Kim
Department of Systems Engineering
Ajou University
Suwon
South Korea

Satchidananda Dehuri
Department of Information and
    Communication Technology
Fakir Mohan University
Balasore, Odisha
India

Gi-Name Wang
Department of Industrial Engineering
Ajou University
Suwon
South Korea

ISSN 2197-6503   ISSN 2197-6511   (electronic)
Studies in Big Data
ISBN 978-3-319-27518-5   ISBN 978-3-319-27520-8   (eBook)
DOI 10.1007/978-3-319-27520-8

Library of Congress Control Number: 2015958330

© Springer International Publishing Switzerland 2016
This work is subject to copyright. All rights are reserved by the Publisher, whether the whole or part
of the material is concerned, specifically the rights of translation, reprinting, reuse of illustrations,
recitation, broadcasting, reproduction on microfilms or in any other physical way, and transmission
or information storage and retrieval, electronic adaptation, computer software, or by similar or dissimilar
methodology now known or hereafter developed.
The use of general descriptive names, registered names, trademarks, service marks, etc. in this
publication does not imply, even in the absence of a specific statement, that such names are exempt from
the relevant protective laws and regulations and therefore free for general use.
The publisher, the authors and the editors are safe to assume that the advice and information in this
book are believed to be true and accurate at the date of publication. Neither the publisher nor the
authors or the editors give a warranty, express or implied, with respect to the material contained herein or
for any errors or omissions that may have been made.

Printed on acid-free paper

This Springer imprint is published by SpringerNature
The registered company is Springer International Publishing AG Switzerland

Bhabani Shankar Prasad Mishra dedicates this work to his parents, wife, and kids.

Satchidananda Dehuri dedicates this work to his wife: Dr. Lopamudra Pradhan, and kids: Rishna Dehuri and Khushyansei Dehuri.

Euiwhan Kim and Gi-Nam Wang dedicate this work to their wives and kids.

Preface

The purpose of this volume entitled: *Techniques and Environments for Big Data Analysis: Parallel, Cloud, and Grid Computing* is to magnetize and sensitize a wide range of readers and researchers in the area of Big data by presenting the recent advances in the fields of Big data analysis, and also the techniques and tools used to analyze it. Further, it can enlighten them on how the expensive fitness evaluation of evolutionary learning can play a vital role in Big data analysis by adopting parallel, grid, and cloud computing environments.

Rapid growth of computational resources produces huge amounts of data to be used in the field of engineering and technology known as Big data. So it is essential to know computational theories and tools which assist humans in extracting knowledge from Big data. The amount of data collected across different areas exceeds human ability to reduce and analyze without the help of the automated machines. There is much knowledge accumulated in the voluminous data. On the other hand, these computational resources can also be used to better understand the data, by performing large-scale evaluations, parameter sweeps, etc. We refer to the overall use of massive on-demand computation (cloud or GPUs) for machine learning as Big Learning. Evolutionary machine learning techniques are perfect candidates for big learning tasks as they have flexibility in knowledge representations, learning paradigms, and their innate parallelism.

To achieve the objectives of Big data, this volume includes ten chapters contributed by promising authors. In Chap. 1, Mishra and Pattanaik have presented an introduction and architecture of Big data. They also present a brief description about Big-table, MapReduce, and Hadoop. Mishra and Sagnika discuss different parallel environments and their architectures in Chap. 2. They have presented a descriptive vision on how to work in different parallel environments.

Dev and Patgiri present an overview on Hadoop distributed file system (HDFS) in Chap. 3. They also evaluate the performance of the Hadoop in different environments by considering several factors. How files less than the block size affects Hadoop's R/W performance and how the time of execution of a job depends on block size and number of reducers are illustrated. They also enumerate some of the challenges of Hadoop.

Mustafi in Chap. 4 draws focus on how Big data challenges can be handled from the data science perspective. The data available for analysis are in different forms in terms of volume, velocity, variety, and veracity. The objective is to resolve some of these real-world problems using natural language processing, where the unstructured data can be transformed into meaningful structured information; and machine learning to get more insights out of the information available or derived. This chapter fairly covers important methodologies where, what, and when to apply. Some open research problems are also shared for the budding data scientists.

In Chap. 5, Panigrahi, Tiwari, Pati, and Das present the development of cyber foraging systems by introducing the concept of cloudlets and the role of cloudlets in cyber foraging systems as well as discuss the working and limitations of cloudlets. This chapter also explores the new architectures where the cloudlets can be helpful in providing Big data solutions in areas with less Internet connectivity and where the user device disruption is high. The chapter then deals with different applications of cloudlets for Big data and focuses on the details of the existing work done with cyber foraging systems to manage different characteristics of Big data.

As the scope of computation is extending across domains where large and complex databases are needed to be dealt with, it has become a very useful approach to subdivide the tasks and to perform them in parallel, which leads to a significant reduction in the processing time. On the other hand, evolutionary algorithms are rapidly gaining popularity to solve intractable problems. However, they are also suffering with an intrinsic problem of expensive fitness evaluation; hence parallelization of evolutionary algorithms proves to be beneficial in solving intensive tasks within a feasible execution time. Therefore to address the aforesaid issues, in Chap. 6 Mishra, Sagnika, and Dehuri present different parallel genetic algorithm models and uses of different Big data mechanisms like MapReduce over parallel GA models.

In Chap. 7, Ghosh and Desarkar present the limitations of general search optimization algorithm for Big data. They also discuss how evolutionary algorithms like GA can be suitable tool for Big data analysis. In Chap. 8, Meena and Ibrahim present how, by using MapReduce programming model, feature selection can be done, when documents are represented as a bag of syntactic phrases. They also explain how ACO can be parallelized using a MapReduce programming model.

In Chap. 9, Mishra and Patel present the key challenges, issues, and applications of grid technologies in the management of Big data.

Finally, we hope that the readers enjoy reading this book, and most importantly, that they learn all new computing paradigms enumerated in this book for analyzing Big data.

Bhabani Shankar Prasad Mishra
Satchidananda Dehuri
Euiwhan Kim
Gi-Nam Wang

Acknowledgments

We express our gratitude to all those who provided support and contributed chapters and allowed us to quote their remarks and work in this book.

We thank Santwana Sagnika for helping us in the process of compilation of this edited volume.

Finally we offer our gratitude and prayer to the Almighty for giving us wisdom and guidance throughout our lives.

Contents

Chapter 1
Introduction to Big Data Analysis

Kiranjit Pattnaik and Bhabani Shankar Prasad Mishra

Abstract The technology in 21st century is evolving at a very high speed and accordingly the data produced are in huge volume and this is where Big data comes into picture. Handling such huge data from different real time sources is a challenge for organizations and in order to resolve this, Apache created a platform named Hadoop whose job is to handle Big data. After that Google created a software framework named Map Reduce which is the main component of Hadoop. Many organizations started experimenting on Big data and created their own small applications to handle their internal business jobs such as creating their own distributed database to work with Big data. This chapter mostly brings into picture the meaning of Big data and its importance and the different frameworks and platforms involved in it, as well as the distributed databases which are used for Big data by different organizations and even the detailed view of the Map Reduce concept which handles Big data in a very efficient manner.

1.1 Introduction

Big data [9, 14] technologies describe a new generation of technologies and architectures, designed to economically extract value from very large volumes of a wide variety of data, by enabling high-velocity capture, discovery, and analysis.

Big data is where the data volume, acquisition velocity, or data representation limits the ability to perform effective analysis using traditional relational approaches or requires the use of significant horizontal scaling for efficient processing. Data sets whose size is beyond the ability of typical database software tools to capture, store, manage, and analyze.

K. Pattnaik (✉) · B.S. Prasad Mishra
School of Computer Engineering, KIIT University, Bhubaneswar, Odisha, India
e-mail: kiranjit.pattnaik99@hotmail.com

B.S. Prasad Mishra
e-mail: mishra.bsp@gmail.com

© Springer International Publishing Switzerland 2016
B.S.P. Mishra et al. (eds.), *Techniques and Environments for Big Data Analysis*,
Studies in Big Data 17, DOI 10.1007/978-3-319-27520-8_1

Big data is a word used to describe data sets so large, so complex or that require such fast processing, that they become very difficult and unmanageable to work with the standard traditional database management or analytical tools. So, massive parallel software is required to handle such data sets running on tens, hundreds, or even thousands of servers.

Big data is the combination of new and old technologies that helps companies to gain actionable insight. Hence it is capable enough to manage large volume of distinct data, at the correct speed, and within the correct time frame for allowing real-time analysis.

Big data [21] is a combination of technologies which include data management that evolved over time. It supports organizations to manipulate, store, and manage huge amounts of data at the correct speed and right time to gain the true insights. The significance of Big data is that the data has to be managed so that it fulfills the business requirement.

Big data [5, 16, 22] is not a secluded solution, yet executing a Big data solution requires the organization to upkeep the management, scalability, and distribution of that data. Therefore, to make use of this significant technology trend it is important to put both a technical and business in action.

The growth of Big data [13, 17] comprise the explosion of video, photos, social media, structured, and unstructured text in addition to the data gathered by abundant sensing devices which includes smart phones too. Capture, search, analysis, storage, sharing, and data visualization are among the many problems accompanying Big data.

Big data is not associated with only transactional data, it acquires data from transactions, interactions may it be social or business and observation as in live feeds. The pictorial representation of it is in Fig. 1.1.

During recent years data has moved from Megabytes to Gigabytes and Terabytes but now data has been moved from Terabytes to Petabytes and Yotabytes and is termed as Big data. At the ERP (Enterprise Resource Planning) level data that is used is of the order Megabytes but at the CRM (Customer Relationship Management) level data is in the order of Gigabytes as it involves technology to organize, marketing, technical support and customer services. At the web level the data is of the order Terabytes it includes web logs, sites, servers, searching, networks, etc. And finally due to huge data from sensors, live videos and images, spatial and GPS coordinates, social networking, business data feeds the order of the data has moved up to Petabytes and even Yotabytes [4]. The pictorial representation of it is shown in Fig. 1.2.

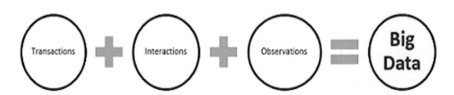

Fig. 1.1 Big data as various forms of data

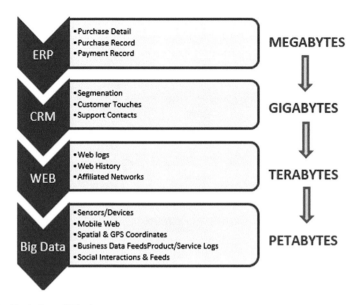

Fig. 1.2 Evolution of Big data

1.2 Aspects of Big Data

Big data consists of different aspects and basically there were only 3 aspects which defined Big data and they are:

- Volume—How much data (Yotabytes, Petabytes, ...)
- Velocity—How fast the data is processed
- Variety—Different types of data

But then gradually some more aspects came into view which would support the previous 3 V's of Big data to further define Big data and those are:

- Veracity—Uncertainty of data (Data Incompleteness)
- Value—Turn Big data into values else useless (Business Perspective)

The whole gamut of Big data is described in Fig. 1.3.

1.3 Objectives of Big Data

The major objective of Big data can be enumerated as below:

1. <u>Cost Reduction</u>:- Organizations following Big data strongly believe that Petabytes of data storage for structured and unstructured data are economically delivered through Big data technologies like Hadoop clusters and the use of distributed

Fig. 1.3 Different aspects
of Big data

database and even by the use of cloud technology. Else previously to handle such
data supercomputers were used which is very much costly and not functional by
every user.

2. Time Reduction:- The common objective of Big data technologies is the reduction
of time. To manage such huge distinct noisy data not only appropriate tools are
required but also accessing them within as much less time as possible. Using the
data and analytics of customer or user past experience to interact with them in
real time is also a part of time reduction.

3. Develop New Big data Tools:- One of the most confusing things an organization
can do with Big data is to use it in developing new product and tools service
based on data. Many of the companies that use such approach are the online
organizations, which have an apparent need to employ data-based products and
services. The best example may be Google and Linked In.

4. Support Management and Business Decisions:- Previously using small data analytics with structured data, certain internal management and business decisions
were carried out. But now a days as much of the data are unstructured and in
large volumes using these previous small data analytics and some Big data tools
in order to improve the analysis of such distinct data and to manage decisions out
of it properly for the benefit of the organizations.

1.4 Big Data Types

- **Structured Data**—It generally refers to the data that has a defined length, type, design and format. It accounts for 20% of the data that is present in the world. Such data can be queried by using Structured Query Language (SQL). The data can be machine generated or human generated.

  - Machine Generated
    Sensor Data—Data involving GPS (Global Positioning System), RFID Tag (Radio Frequency ID), Medical Devices, Smart meters.
    Web Log Data—Data from networks, servers, application and prediction of security breaches and also data dealing with service level agreements.
    Point-of-Sale Data—Cashier swiping the bar code of any product
    Financial Data—Stock Trading data (company symbol, currency value)
  - Human Generated
    Input Data—Input by user such as name, address, age, income, etc.
    Click-Stream Data—Data generated whenever click is done on a link to determine customer behavior and buying patterns
    Gaming Related Data—Every move in a game can be recorded.

- **Unstructured Data**—It generally refers to the data having variable length, type, design or format. Most of the data in this world are unstructured. The data can be machine generated or human generated.

  - Machine Generated
    Satellite Image—Weather Data and Satellite images(Google Earth)
    Scientific Data—Seismic Imagery, Atmospheric Data, High Energy Physics
    Photographs and Videos—Security, Surveillance and Traffic videos
    Radar/Sonar Data—Vehicular, Meteorological and Oceanographic seismic profiles
  - Human Generated
    Text Internal to your Memory—Text within documents, logs, survey results, emails, etc.
    Social Media Data—Social Media Platforms such as YouTube, Facebook, Twitter, Linked In, etc.
    Mobile Data—Text Messages, Location Information
    Website Content—Site delivering unstructured content.

- **Semi-Structured Data**—It is the intersection of structured and unstructured data. It doesn't necessarily confront to a fixed schema but may be self-defining and have simple pairs of label/value.

Different data types of Big data with examples are shown in Fig. 1.4.

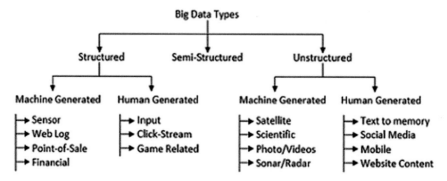

Fig. 1.4 Different data types of Big data

1.5 Classification of Big Data

Broadly Big data can be classified based on data sources, data frequency, data consumers, content format, data stores, data staging, and data processing. The summarized view of classification of Big data with its respective classes is shown in Fig. 1.5.

- Data Sources—Machine Generated, Human Generated, Web and Social Media, Internal Data Sources, Transaction Data, Bio-metric Data, Sensing
- Data Frequency—On-demand Feeds, Continuous Feeds, Real Time Feeds, Time Series
- Data Consumers—Human, Business Process, Enterprise Applications, Data Repositories
- Content Format—Structured, Semi-structured, Unstructured
- Data Stores—Document Oriented, Column Oriented, Graph Based, Key Valued
- Data Staging—Cleaning, Normalization, Transform
- Data Processing—Predictive Analysis, Query and Reporting, Analytical (Speech Analytics, Statistical Algorithms Text Analytics, Social Network Analytics, Location Based Analytics)
- Analysis Type—Batch, Real Time.

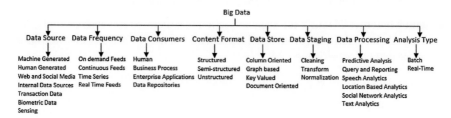

Fig. 1.5 Classification of Big data

1.6   Architecture of Big Data

Big data [1, 15, 18] follows a layered architecture. Different layers present in Big data architecture are explained below. The layered view of the architecture of Big data is shown in Fig. 1.6.

- Data Source Layer—It is the layer which provides input through the relational database and even the structured, unstructured and semi-structured types of data.
- Ingestion Layer—It is the layer to import and process the data, separate noise from the data and store it for further use by tools such as Flume, Kafka, Storm.
- Visualization Layer—It is the layer to gain insights and improve the ability to look at different aspects of Big data by the data scientists and analysts by the Hadoop Administration, Data Analyst IDE/SDK, Visualization tools.
- Analytics Engine Layer—In this layer different approaches are adopted to handle Big data by Statistical Analytics, Text Analytics, Search Engine and Real Time Engines.
- Hadoop Platform Management Layer—In this layer tools like Pig, Hive, Oozie, Map Reduce, Zookeeper, Impala, Spark and Shark are used to manage the huge volume of data using HDFS.
- Hadoop Storage Layer—In this layer the data is stored at low cost using NoSQL Database and Hadoop Distributed File System (HDFS) for high velocity distributed processing of algorithms.
- Data Warehouse—It manages the RDBMS based data in a centralized environment and even consists of some analytic appliances
- Memory—The layer in which nodes are present where the data is stored or processed in Rack, Disk and CPU.
- Hadoop Infrastructure Layer—It consists of bare Metal Clustered Workstations, Virtualized Cloud Service and it supports the Hadoop Storage Layer. In this layer the data is stored in many different locations in a distributed model and linked together. Its main aim is to check the redundancy of the huge data.
- Security Layer—Tasks such as Data Encryption, Data Access, Threat Detection, Nodes Authentication and tools such as Big-IP, Apache Sentry, Apache Knox are used to maintain privacy and to meet the compliance requirements for proper authorization and authentication.
- Monitoring Layer—In this layer the distributed clusters are monitored using tools like Chukwa, Hue, Ganglia, Open TSBD, Nagios as such huge data is handled.

1.7   Big Table, Map Reduce and Hadoop

The innovations of Map Reduce, Hadoop and Big Table provide a spark on managing huge data. These technologies help in addressing the basic problems like the capability to process massive amounts of data efficiently and timely and cost effectiveness.

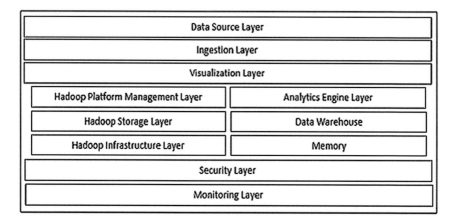

Fig. 1.6 Architecture of Big data

1.7.1 Hadoop

Hadoop [9, 10, 20] is a software framework managed by Apache and derived from Big Table and Map Reduce and created by Doug Cutting and was named after his son's toy elephant. It permits applications to run on huge clusters of commodity hardware based on Map Reduce. It is designed to parallelize the data processing across different computing nodes to speed up computations and hide the latency. There exists two major components of Hadoop which is an immensely scalable distributed file system that can support yotabytes and petabytes of data and an extremely scalable Map Reduce engine which computes the results in batches.

Hadoop [9, 11, 22] is a universal processing framework designed to execute queries and other batch type read operations on huge data sets that scale upto hundreds of terabytes to petabytes in size. The Hadoop platform comprises a consistent distributed file system called the Hadoop Distributed File System (HDFS) and also a parallel data processing engine called the Hadoop Map Reduce.

The Map Reduce and HDFS [8, 12, 19] together provides a software framework for processing huge data in parallel on large clusters of commodity hardware (possibly thousands of nodes) in a fault-tolerant and reliable manner. The architecture of Hadoop Map Reduce is shown in Fig. 1.7.

1.7.1.1 Characteristics of Hadoop

1. Fault Tolerance
2. Open Data Format
3. Flexible Schema
4. Queryable Database

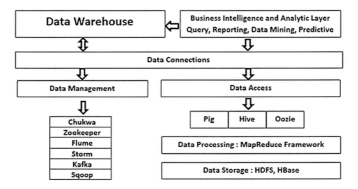

Fig. 1.7 Architecture of Hadoop Map Reduce

5. Archival Store
6. Never Delete Data
7. Distributed File System

1.7.2   Big Table

Big Table was developed by Google. It is a distributed storage system aimed at managing highly accessible structured data. Data is ordered into tables with columns and rows. It is persistent, distributed and sparse multidimensional sorted map unlike traditional relational database. It stores large volumes of data through commodity servers.

1.7.2.1   Data Access

Data access usually refers to accessing and handling the stored data from any database to any external devices. In this case accessing such huge data, some of the tools that are required for Data Access are mentioned below and its summarized view is shown pictorially in Fig. 1.8.

- Pig—It is a project of Apache Software Foundation. It is a platform used for the analysis of huge data sets and also for making the data flows for the ETL (extract, transform, and load) processing. It is also a script based interactive execution environment which supports Pig language.
- Hive—It is a project of Apache Software Foundation. It is an SQL based data warehouse system that assists ad hoc queries, data summarization and analysis of huge data sets stored in Hadoop friendly file system.
- Oozie—It is a Apache Hadoop Map Reduce work-flow scheduler and manager.

Fig. 1.8 Different tools for
data access in Big data

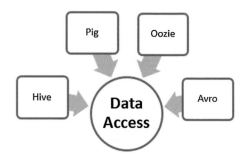

1.7.2.2   Real-Time Processing

It refers to a processing in which the system imports and updates the data and at the same time produces results which supports further processing or acts as input for the next phase. In terms of Big data, for handling such huge data in real-time some tools has been designed which are explained below and are shown pictorially in Fig. 1.9.

- Impala—It is an open source by Cloudera. It is a new query engine that bypasses MapReduce for very fast queries over data sets in HDFS. It uses HiveQL as the query language.
- Spark and Shark—It is a product of Apache Software Foundation. It is a newer distributed computing framework that exploits sophisticated in-memory data caching to improve many common data operations, sometimes by multiples of 10x.
- Infosphere—It is a platform used by IBM to perform difficult analytics of complex/heterogeneous data types. Its Streams can support all types of data. It can even perform look ahead analysis of regularly generated data and real time, using pattern/correlation analysis and decomposition, digital filtering as well as geospatial analysis.

1.7.2.3   Ingestion and Streaming

Ingestion and Streaming deals with the flow and speed of massive unstructured data, importing and storing such huge data for later use. There are many tools which

Fig. 1.9 Different tools for
real-time processing in Big
data

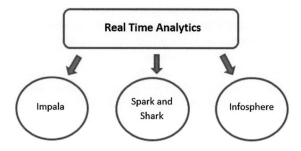

handles these unstructured data and are explained below. The summarized view is shown in Fig. 1.10.

- Flume—It is a product of Apache Software Foundation. It is a distributed system for collecting and aggregating log data from external sources and writing it to the HDFS. It is very reliable, simple, flexible and highly available. It is also an inbuilt programming model based on streaming data flows.
- Chukwa—It is an Apache Hadoop sub-project that links that gap between Map Reduce and log handling. It provides a scalable distributed system for analysis and monitoring of the log based data.
- Sqoop—It a product of Apache Software Foundation. It is a tool designed for efficiently transferring bulk data between Apache Hadoop and structured data stores such as relational databases.
- Kafka—It is a project of Apache Software Foundation. It is a distributed messaging system which is designed to provide high throughput tenacious messaging that's scalable and also allow parallel loads of data into Hadoop.
- Akka—It is a distributed messaging system with actors. It is an open source toolkit and runtime for building highly concurrent, distributed and resilient message driven applications on JVM.
- Zookeeper—It is a software project of Apache Software Foundation. It is a centralized service for maintaining configuration information, naming and providing distributed synchronization.

Fig. 1.10 Different tools for ingestion and streaming

- Storm—It is a project of Apache Software Foundation. It is an event processing, batch mode processing and a general purpose system that is growing in acceptance for managing the gap in Hadoop over massive data sets.

1.7.2.4   Machine Learning in Hadoop

Machine learning itself is defined as a scientific discipline which studies or analyses the algorithms associated with the data and learns from it which in turn gradually helps for self evolving instead of just following some algorithm directly. In this case as Big data is handled in Hadoop platform some of the tools used for machine learning in case of Big data in Hadoop are mentioned below and the summarized view of the tools is shown in Fig. 1.11.

- Mahout—It is tool by Apache Software Foundation. Its goal is to build scalable machine learning libraries. Mahout's core algorithms for clustering, classification, and batch-based collaborative filtering are implemented on top of Apache Hadoop using the Map Reduce paradigm.
- Giraph—It is a system developed by Apache Software Foundation. It is an iterative graph processing system on top of Hadoop built for high scalability.
- RHipe—R+Hadoop (Statistical Computing). R is a software language for carrying out complicated statistical analyses. It includes routines for data summary and exploration, graphical presentation and data modeling. R+Hadoop is also known as Revolutionary Analytics.
- Clojure—It is a dynamic programming language based on LISP (created by Rich Hickey) that uses the Java Virtual Machine (JVM). It is well suited for parallel data processing.

Fig. 1.11 Different machine learning tools in Hadoop

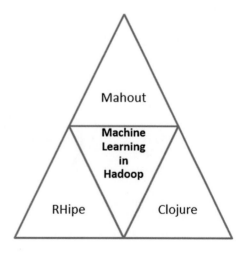

1.7.2.5 NoSQL

It is a term used for a wide class of database management systems that ease some of the traditional design constraints of RDBMS (Relational Database Management Systems) [22] in order to meet the goals of more scalable, cost-effective, consistency and flexibility for data structures that are not suitable in the relational model, such as the large graphs and key-value data. Some of the NoSQL databases are explained below and the summarized view of the databases is shown in Fig. 1.12.

- HBase—It is a project by Apache Software Foundation under Apache Software License v2.0. It is a column oriented, distributed database, where each cell has a organized number of previous values that is retained. It supports ability of Big table on Hadoop. It supports SQL queries with high latency using Hive.
- Cassandra—It is an open source project by Apache Software Foundation. It is the most popular NoSQL database for very huge data sets. It is a clustered, key-valued database that uses the column-oriented storage and redundant storage for accessibility in both read/write performance and data sizes.
- MangoDB—It is the name of the project for 'hu-mango-us database' system. A company named 10gen maintains it as open source and is available free under GNU AGPL v3.0 license. It is a document-oriented NoSQL database with each record a JSON document. It has a rich query language which is javascript based that exploits the inherent structure of JSON. It supports sharding for better resilience and scalability. It is also termed as a grid based file system (GridFS) which enables the storage of large objects a multiple documents.
- DynamoDB—It is Amazons highly available, scalable, key-valued NoSQL data-base. It is one of the earliest NoSQL databases and it affects the design of other NoSQL databases, such as Cassandra.
- CouchDB—It is an open source project by Apache Software Foundation under Apache Software License v2.0. It is a key valued NoSQL database which suits mobile applications whose copy of a data set resides on many devices in which changes are performed on a copy and are synchronized with the availability of the connectivity. It can continue to operate in areas with less network connectivity as it is resistant to network dropouts.
- Redis—It is also known as Remote Dictionary Server. It was developed by Salva-tore Sanfilippo. It is a key value store which supports fundamental data structures as values, including lists, hash maps, strings, sets, and sorted sets. It acts as a data structure server.
- Datomic—It is developed by Rich Hickey, Stuart Halloway, Justin Gehtland. It is an implementation on datalog on Clojure. It is a new type of NoSQL landscape with an exclusive data model that evokes the state of the database at all points in the past, making a reconstruction of state and events insignificant. Its deployments are highly distributed, available and elastic.
- Riak—It is an open source, distributed database. It is master less fault tolerant, key valued and distributed NoSQL database considered for large scale deployments in

Fig. 1.12 Different NoSQL
databases

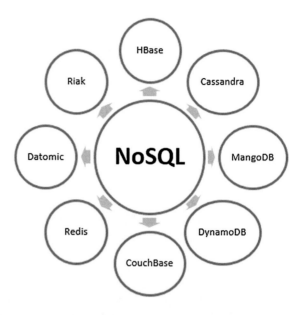

hosted environments or cloud with no single points of failure. It is robust against
multiple nodes failure and nodes can be easily removed or added.

1.7.3   Map Reduce

Map Reduce [2, 14, 23] is designed by Google as a framework which can execute
a set of functions efficiently against huge volume of data in batch mode. The 'Map'
module divides the problem or tasks across large number of systems and handles
the tasks by balancing the load and by managing recovery from failures. After the
completion of the 'Map' phase, another function 'Reduce' is called which combines
all the elements back together as per required function to provide a result.

Map Reduce is the heart of Hadoop. It is this programming paradigm that allows
for massive scalability across hundreds or thousands of servers in a Hadoop cluster.
The Hadoop programs perform two distinct tasks of Map Reduce. The first is the
Map task, where a set of data is taken as input and then converted into another set of
data for further processing, where each individual component is broken down into
key/value pairs. The second is the Reduce task whose input is the output produced by
the Map task and then it combines those key/value pairs into another set of key/value
pairs. The reduce task is always performed after the Map task.

Map Reduce [6, 7] is a framework which processes parallelizable problems
through large sets of data with large number of computers (nodes), collectively
termed as a cluster (if all nodes use similar hardware and are on the same local
network) or a grid (if the nodes are shared across administrative and geographic

distributed systems, and using more heterogeneous hardware). Processing can occur on data stored either in a file system (unstructured) or in a database (structured). Map Reduce takes the advantage of data locality, processing on or near the storage devices to reduce the distance over which it should be transmitted.

- Map—The map function is applied by each worker node to the local data, and the output is written to a temporary storage. A master node checks that for duplicate copies of input data, only one of them is processed.
- Shuffle and Sort—Based on the output of the map function the worker node redistributes the data such that all data fitting to one key is placed on the same worker node.
- Reduce—Each group of output data per key is processed by the worker nodes in parallel.

1.7.3.1   Steps of Map Reduce

1. Modify the Map input—The Map Reduce framework defines 'Map' processors, allocates the input key/value pair K1/V1 to each individual processor to perform its work on, and provides that particular processor with all the input data associated with that key/value.
2. Execute the user-defined Map code—Map task is run exactly once for each key/value pair K1/V1, to produce output structured by key/values pair K2/V2.
3. Shuffle and Sort the Map output—The Map Reduce framework designates Reduce processors, assigns the key/value pair K2/V2 to each processor for performing its required function, and provides that processor with all the data generated by the Map task related with that key/value pair.

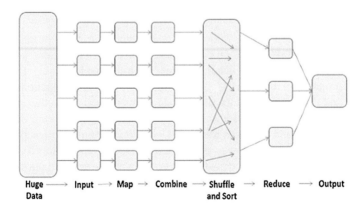

Fig. 1.13 Different steps involved in Map Reduce

4. Execute the user-defined Reduce code—Reduce code is executed precisely once for each key K2 and a list of values 'list(V2)' produced by the Map step to produce another key/value pair K3/V3.
5. Final output—The Map Reduce framework collects all the Reduce output, and sorts it by key/value K3/V3 to produce the final result. The pictorial representation of the steps involved in Map reduce in shown in Fig. 1.13.

1.7.3.2 Overview of Map Reduce

- Consists of two functions operating on key/value pairs
  - Input:- key/value pairs
  - Output:- key/value pairs
- Map Function
  - Performs filtering and sorting
  - Map(key 1, value 1) → (key 2, value 2)
- Reduce Function
  - Performs summary operations on Map step results
  - Reduce(key 2, list(value 2)) → (key 3, value 3)
- Partition Function
  - Hash(key) % (No.of Reducers)

1.7.3.3 Characteristics of Map Reduce

1. Scalable—Map Reduce usually handle large data and it has the ability to process such huge data in a capable manner.
2. Fault Tolerant—Map reduce works with nodes and if during its process there is a node failure then the whole process doesn't stop rather the current work of the faulty node stops and it gets handled by other working nodes.
3. Batch Computation in parallel—All the mapping and reduce functions are pre-defined and due to the presence of multiple nodes in map and reduce phase the task is automatically divided into the total number of nodes in each of the map and reduce phase and produces the output without any human interaction.
4. Graceful Degradation—Because of Map Reduce fault tolerant nature it can handle any faulty situation and automatically tries to complete the task without any human help.
5. Generic Programming—As previously known Map Reduce has two phases i.e. the Map phase and Reduce phase but the function that is to be written in the map and reduce part can be user defined too. So according to the task coding is done such that the required result is found out by Map Reduce processing.

Fig. 1.14 Map Reduce data
flow

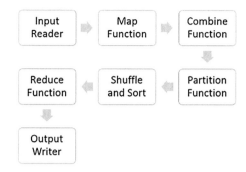

1.7.3.4   Map Reduce Data Flow

1. Input Reader—It divides the input into appropriate size 'splits' and each individual is to each map function by the framework.
2. Map Function—A sequence of key/value pair is taken and each individual is processed and produces zero or more output key/value pairs.
3. Combine Function—It combines all the similar key/value pair before the Shuffle and Sort and Reduce phase. It is also a type of reduce phase but occurs before shuffle and sort phase. It is necessary for arithmetic type data operations.
4. Partition Function—The output of each map function is assigned to a particular reducer by the partition function for sharding purposes. The key and the number of reducers is given to this function which then returns the index of the preferred reducer.
5. Shuffle and Sort—The input for each Reduce is retrieved from the local memory where the map function was executed and then sorted using the compare function.
6. Reduce Function—The reduce function for each individual intermediate unique key is called by the framework in a sorted manner. It can then repeat through the values that are associated with that key and finally produce zero or more outputs.
7. Output Writer—It writes the output produced by the reduce function to the steady distributed file system storage.

Overview of the flow of the data in the Map Reduce is shown in Fig. 1.14.

1.7.4   Different Platforms Available to Handle Big Data

- Root

  - C++ Framework for Petabyte Storage, Statistical Analysis and Visualization.
  - System for large scale data analysis and data mining.

- MAD

  - Magnetic, Agile, Deep

- Used to get the organizations data into the warehouse as soon as possible.
- Cleaning and Integration of data is staged intelligently and very efficiently.

- Starfish

  - Self tuning system for data analytics build on Hadoop
  - Adaptive to user needs and system workloads to provide good performance automatically without user interfering in Hadoop usage.
  - Automatically converts workloads in higher level languages to Lastword (A language used by Starfish to optimize different workloads that run on same Hadoop cluster)

- R

  - It is a software environment aimed at Statistical Computing and Graphics.
  - R+Hadoop = Revolutionary Analytics
  - An integrated suit of software facilities for data manipulation, data handling and storage facility, calculation and graphical display.

1.7.5 Different Frameworks for Big Data

- Avro—It is a Data serialization system developed by Apache Software Foundations. It provides remote procedure cal (RPC), fast, compact, binary data format and also a container file which stores determined data.
- Trevni—It is a framework that works with column file format.
- Protobuf—It is developed by Google and is a standard serialization library. It is similar to Apache Thrift.
- Parquet—It is a columnar storage format for Hadoop build by Apache Software Foundation, regardless of any data model or data processing framework or any programming language.
- Kiji—It is an open source project for building real-time applications of Big data on Apache HBase.
- Elephant Bird—It is a framework used for serializers and compression codes for Hadoop.
- Summing Bird—It is a framework for large scale data processing which enables developers to execute their codes in batch or stream mode. It is an open source language for online Map Reduce on Storm or Scalding
- Apache Crunch—It is an efficient, simple Map Reduce pipelines for Spark and Hadoop developed by Apache Software Foundation.
- Cask—It is a cloud based application platform for distributed data and was formerly known as Continuuity.
- MrUnit—It is a unit testing framework by Apache Software Foundations for Java Map Reduce.
- PigUnit—It is a xUnit framework which is used for testing Pig scripts.

1.8 Conclusion

Driven by the real world applications and scenarios, Big data has been very challenging and fascinating. The need to process such huge quantities of data has never been greater. As most of the data present includes noise in it and to extract the required data from such huge space is really challenging. Different tools and frameworks have been designed that can solve the problem of huge inconsistent data by Big data. In this age of social technology data if not handled properly would give no sense without the use of Big data. To handle yotabytes and petabytes of data some Big data analytic tools has been developed like the Map Reduce over Hadoop and a distributed storage system to handle and store such huge data like the HDFS (Hadoop Distributed File System). Big data promises to help organizations to understand their customers, clients and marketplace better and hence leading to better business decisions and analysis.

But there exists some issues/challenges in Big data. Some of them are elaborated below:

- Understanding and prioritizing the huge data from the junk that enters the enterprises.
- 90 % of data reflects noise and the task is heavy to sort and filter the knowledge out of such noisy data.
- The security of such huge data and its abstraction.
- Cloud computing and virtualization usage further confuses the decision to handle and host Big data solutions outside the initiative.
- Costly to archive such huge amounts of data.
- Unavailability of skills to work with Big data.

References

1. 10 of the most popular Big data tools for developers. http://www.cbronline.com/news/big--data/analytics/10--of--the--most--popular--big--data--tools--for--developers--4570483.
2. 50 Top open source tools for Big data—Datamation. http://www.datamation.com/data--center/50--top--open--source--tools--for--big--data--1.html.
3. Apache hadoop. http://hadoop.apache.org/.
4. Big data: all you need to know—ZDNet. http://www.zdnet.com/article/big-data-all-you-need-to-know/.
5. Davenport, T. H., & Dyche, J. (2013). Big data in big companies. *International Institute for Analytics*.
6. Dean, J., & Ghemawat, S. (2008). MapReduce: Simplified data processing on large clusters. *Communications of the ACM*, *51*(1), 107–113.
7. DZone. https://dzone.com/articles/big--data--beyond--mapreduce.
8. Hadoop Fundamentals I. http://bigdatauniversity.com/bdu--wp/bdu--course/hadoop--fundamentals--i--version--3/.
9. http://en.wikipedia.org/wiki/Big_data.
10. http://wiki.apache.org/hadoop.
11. http://www.tutorialspoint.com/hadoop/hadoop_big_data_overview.htm.

12. http://www.zettaset.com/index.php/info--center/what--is--big--data/.
13. Hu, H., et. al. (2014). Toward scalable systems for Big data analytics: A technology tutorial. *Access IEEE, 2*, 652–687.
14. Hurwitz, J., et. al. (2013). *Big data for dummies*. Wiley, Hoboken.
15. Jorgensen, A., et. al. (2013). *Microsoft Big data solutions*. Wiley, Indianapolis.
16. Know Your Big data—in 10 Minutes! (2012). *HappiestMinds Technologies*.
17. Purcell, B. The emergence of big data technology and analytics. *Journal of Technology Research,* Holy Family University, pp. 1–6.
18. Sawant, N., & Shah, H. (2013). Big data application architecture Q & A.
19. Shvachko, K., Kuang, H., Radia, S., & Chansler, R. (2010). The Hadoop distributed file system. In *IEEE 26th Symposium on Mass Storage Systems and Technologies (MSST), Vol. 1, No. 10,* pp. 3–7.
20. What are big data techniques and why do you need them?—GCN. http://gcn.com/microsites/2012/snapshot-managing-big-data/01-big-data-techniques.aspx.
21. What is Big data? A Webopedia definition. http://www.webopedia.com/TERM/B/big_data.html.
22. www.knowbigdata.com.
23. Yang, X., & Sun, J. (2011). An analytical performance model of MapReduce. In *IEEE Conference on Cloud Computing and Intelligence Systems (CCIS),* pp. 306–310.

Chapter 2
Parallel Environments

Bhabani Shankar Prasad Mishra and Santwana Sagnika

Abstract To cater to the increasing demand for high-speed and data-intensive computing in the current scenario, a widely followed approach has been the development of parallel computing techniques, which enables simultaneous processing of a huge amount of data. By implementing such mechanisms, the total execution time is greatly reduced, and the available resources are utilized in a most efficient manner. This chapter provides a bird's-eye view of the various parallel processing models available in literature and their distinctive features.

Keywords Parallel processing · Message passing · Object-oriented · Shared memory · Partitioned global address space · Algorithm skeleton · Bulk synchronous programming

2.1 Introduction

The increasing dependence on automated systems has widened the scope of computers to work in various fields and solve large complex problems. This involves dealing with large datasets. Even so, performing the necessary tasks in a minimum time-frame is also of paramount importance. Hence, a solution for providing high speed computation of huge amount of data comes in the form of parallel processing. The advent of parallel processing has expanded the application of computers to areas like weather forecast, image processing, military applications, aviation, etc.

To provide parallelism, a possible approach can be to increase the number and power of processors. This may not always be feasible, considering the high costs involved. Practical solutions for parallelism can be implemented by following certain models, as well as programming tools that match the underlying parallel models. Parallelism is achieved at the basic level by two mechanisms data-parallelism and

B.S.P. Mishra (✉) · S. Sagnika
School of Computer Engineering, KIIT University, Bhubaneswar, Odisha, India
e-mail: mishra.bsp@gmail.com

S. Sagnika
e-mail: santwana.sagnika@gmail.com

© Springer International Publishing Switzerland 2016
B.S.P. Mishra et al. (eds.), *Techniques and Environments for Big Data Analysis*,
Studies in Big Data 17, DOI 10.1007/978-3-319-27520-8_2

control parallelism. Data parallelism involves dividing the data into smaller sets and running the same set of instructions on all of them over different processors. On the other hand, in control parallelism the instructions are divided over various processors that execute concurrently. The parallel programming tools rely on either or both mechanisms [12, 13].

For parallel processing, the hardware requirements can be described as follows.

Multi-core systems—In such a system, a single chip contains a number of cores which can operate independently. The memory can be shared. The throughput of the system can be multiplied by the number of processors.

Multi-processor systems—In this case, multiple processors collaborate to perform a parallel task. This can be either (i) Symmetric Multi-Processor (SMP) and (ii) Non-Uniform Memory Access (NUMA). The memory can either be shared or distributed. Communication occurs through high-speed bus. Every processor has a control unit.

Cluster systems—A cluster represents a group of nodes (desktops, servers and workstations) connected over a network. Every node is a self-sufficient system and is able to parallelize computation as well as access to memory. A cluster provides good scalability, and hence, better bandwidth and storage capacity.

Parallel programming models comprise of a combination of mechanisms that implement parallelism as suited to the system that they run on. A parallel programming model and its analogous cost model together forms a parallel model of computation. The parallel programming model simulates a virtual parallel machine that performs basic operations and also includes a memory model that controls visibility of memory to different modules of the parallel system. This model encompasses the necessary frameworks or languages that implement the said model. Besides, the parallel cost model associates specific costs to the various basic operations and also provides mechanisms to estimate the overall costs for the complete task executing in parallel [4].

2.2   Commonly Used Models

The most popular parallel models are discussed here.

2.2.1   Message Passing Models

In this model, a set of tasks perform computation using their local memory. Multiple tasks reside on the same or across multiple machines. Some amount of programming is needed to use multiple nodes for a job to attain faster execution. The various processors executing a particular job can communicate over the network. For writing such programs, certain language-free protocols are used, which have their own implementations using various languages. These are added as routines to the programs. Figure 2.1 shows a sample message passing model architecture [2].

Fig. 2.1 Message passing
model architecture

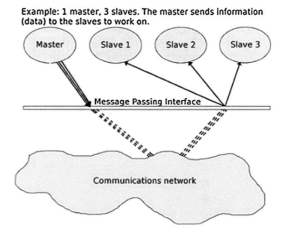

Example: 1 master, 3 slaves. The master sends information
(data) to the slaves to work on.

Various available message passing models are as described below [9, 14].

Model	Features
Message Passing Interface (MPI)	• Language-free communication protocol • Group of processes connected by a Communicator • Performs point-to-point communication using MPI_SEND and MPI_RECV • Runs in three modes—ready, standard and synchronous • Collective/group communication via MPI_Bcast • Uses pre-defined datatypes like MPI_CHAR, MPI_INT, etc • Doesn't define fault tolerance
Parallel Virtual Machine (PVM)	• Appropriate for coarse-grained parallelism • Library modules provide control, locks, message passing and broadcasting • Logical names assigned to executable files • Finds versions suitable for compilation on a specific machine
ClusterM	• Works on heterogeneous environments • Partitions programs and assigns them to parallel machines • Maintains links using trees • Forms clusters of correlated nodes • Concurrent executions of same level clusters is possible
MATLAB MPI	• MPI implemented through MATLAB scripts • Performs coarse-grained parallelism • Follows MPI standards • Realizes global semantics of data distribution & communication

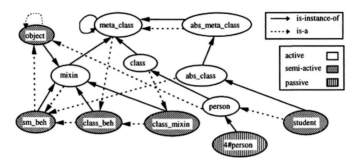

Fig. 2.2 Parallel object-oriented model architecture

2.2.2 Parallel Object-Oriented Models

This model provides software engineering-based methods and abstractions for designing applications. Objects are designed encapsulating their internal states. This model can be considered as a group of shared objects, just like a shared data model. It is considered as a potential technique for parallel models, even though the current inclination of programmers towards traditional programming languages can be an obstacle for its wide acceptance. Some compatible environments like Mentat and PC++ have been discussed in existing literature [17]. Figure 2.2 shows a parallel object-oriented model architecture.

Various parallel object-oriented models are described below [10, 13].

Model	Features
PC++	• Exploits medium-to-coarse grained parallelism • Messages are used for communication • Compiler provides synchronization
Jade	• Works on coarse-grained parallelism • Dynamic extraction from a serial program • Analysis of dependencies and scheduling done at runtime
p-CORBA	• Provides concurrency to standard CORBA architecture • Adds load-balancing feature
Mentat	• Supports messaging between objects • Auto-management by the compiler • Medium-to-coarse grained parallelism
ABCL/1	• Communication though message queues • Validity of messages defined by script construct • Based on the Actor model
Charm++	• Distributes parallelization task between programmer and system • Programmer indicates parallelism • System performs workload distribution • Supports both data and control parallelism • Implements medium-to-coarse grained parallelism

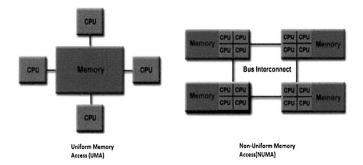

Fig. 2.3 Shared memory programming model architecture

2.2.3   *Shared Memory Programming Models*

In this model, the program is made to run on one or more number of processors, that share a common memory space that can be used to write and read data in asynchronous manner. Concurrent access control mechanisms are required to be implemented. It can utilize Unified Memory Access (UMA) and Non-Uniform Memory Access (NUMA). Data ownership norms are not strict. Program development can be made simple but locality management issues become difficult [1, 5]. Figure 2.3 shows the shared memory programming architecture.

Various shared memory models are discussed below [13, 16].

Model	Features
Threads	• Threads provide faster switching • POSIX threads (Pthreads) provide programming standards • Provide synchronization, control and scheduling • Communicate through the global memory
ADA 9X	• Stored shared code and data in partitions • Partitions interact through RPCs
Delirium	• Coordinating language that uses FORTRAN and C • Medium-to-coarse grained parallelism support • Access to local data faster than remote data
Global Array	• Asynchronous access to blocks of multi-dimensional arrays • Base for current Global Address Space (GAS) languages • Access to local data faster than remote data
Open MP	• Uses fork-join multi-threading • Compiler and library functions control the working • Task allocation through work-sharing constructs
High-performance FORTRAN	• Library functions handle parallelism • Medium-to-fine grained parallelism

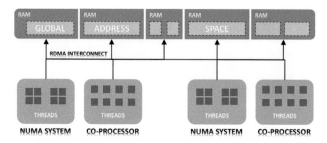

Fig. 2.4 PGAS model architecture

2.2.4 Partitioned Global Address Space Models

Partitioned Global Address Space (PGAS) assumes a logically partitioned memory space in which every partitioned is allocated as local to a processor. The sections of the shared memory space exploit locality of reference and have preference towards particular processes. This model tries to merge SPMD programming for distributed memory systems and the data referential semantics for shared memory systems. This helps to model the hardware-specific locality of data [6, 8]. Figure 2.4 shows the PGAS model architecture. Various PGAS models are discussed as follows [13, 17].

Model	Features
Unified Parallel C (UPC)	• Uses Single Program Multiple Data model
	• Parallelism amount fixed from startup
	• Runs on both shared and distributed memory hardware
X10	• Divides computations into places holding data and related activities
	• Supports user-defined struct types
	• Constrained structure for object-oriented programming
Chapel	• Multi-threaded model
	• High-level abstraction for parallelism
	• Generic programming features and support for code reuse
	• Portable model suitable for clusters

2.2.5 Algorithm Skeleton Models

Algorithmic skeletons represent a higher level programming approach that uses general programming patterns that abstract the complexity of parallel systems, by building complex patterns from more basic ones. The skeleton optimizes the structure to achieve good performance and better adaptation to external environments. After definition of the structure, the programmer implements the functional aspects by writing what is known as *muscle codes*. Figure 2.5 shows an algorithm skeleton architecture [3].

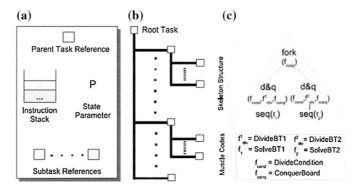

Fig. 2.5 Algorithm skeleton model architecture. **a** Task definition, **b** task tree, **c** N-Queens skeleton

Different algorithm skeleton models are described as follows.

Model	Features
JaSkel	• Framework based on Java • Implementable on grid and cluster structures • Provides concurrent, sequential and dynamic skeletons
Calcium	• Provides a type system for nesting of skeletons • Performance tuning model for debugging • Transparent file access model
ASSIST	• Sequential and parallel models • Performance portability • High-level programmable software • High reusability
Skeleton-BAsed Scientific COmponents (SBASCO)	• Suitable for numerical applications • Customized component language
Higher-Order Divide and Conquer (HDC)	• Polymorphic functions linkable to skeletons • Estimates ratio of tasks to processors

2.2.6 Bulk Synchronous Programming Models

A BSP model consists of a distributed memory architecture, an algorithmic framework, a cost calculation model and a barrier synchronization mechanism. In this model, a master co-ordinates a group of workers. The execution takes place in a lock-step manner. Each worker reads its allocated data from a queue and performs specific processing on it. Then it sends its own result to the communication channel. This process is repeated till there are no more active workers remaining [15]. Figure 2.6 shows a BSP model architecture. Some BSP models are discussed as follows [7, 11, 18].

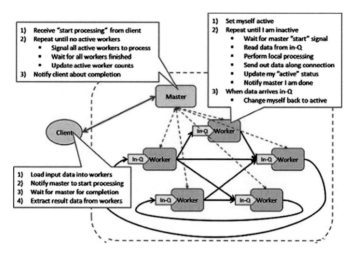

Fig. 2.6 BSP model architecture

Model	Features
Bulk Synchronous Parallel ML (BSML)	• Follows data-parallel mechanism • Collective communications • Avoids deadlocks • Based on primitives accessing physical parameters of the machine
BSPLib	• Small library with 19 operations • Provides data parallelism in Single Program Multiple Data manner • Follows Direct Remote Memory Access
Apache Hama	• Runs on very large datasets in Hadoop Distributed File System • Easier and flexible than traditional message passing models • Speeds up the iteration processes
Pregel	• Works on parallelization of graph algorithms • Vertex-centric approach with message interchange • Synchronous and deadlock-free
BSPonMPI	• Runs on MPI enabled platforms • Uses communication procedures of MPI • Uses request and delivery buffers for communication
Multicore BSP	• Designed for C programming on ANSI C99 standard • Extends BSPLib by adding high-performance primitives • Thread-based parallelization • Suited for shared memory structure

2.3 Conclusion

Parallel computation is an efficient mechanism to implement feasible solutions for complicated and large problems. Such problems are categorized according to the way they communicate data, i.e. either by accessing a shared memory or by passing messages. The selection of the mechanism depends upon the implementation architecture as well as the features of the problem. A suitable language needs to be chosen to specify the partitioning details. The design of the parallel algorithm needs to be done according the environment, i.e. hardware and software on which it needs to be implemented. Parallelism has gained popularity in recent times due to the increasing availability of appropriate machines. This chapter has discussed various characteristics of parallel algorithms and the tools available for their implementation.

References

1. A parallel programming model. http://www.mcs.anl.gov/~itf/dbpp/text/node9.html
2. Beginner guide—www.hpc2n.umu.se. https://www.hpc2n.umu.se/support/beginners_guide
3. Caromel, D., & Leyton, M. (2007). Fine tuning algorithmic skeletons. In A.-M. Kermarrec, L. Bougé, & T. Priol (Eds.), *Euro-Par 2007 Parallel Processing* (Vol. 4641, pp. 72–81)., LNCS Berlin: Springer.
4. Chandrasekaran, K. Analysis of different parallel programming models. Indiana University Bloomington.
5. Demmel, J., & Yelick, K. Shared memory programming: Threads and OpenMP. http://www.cs.berkeley.edu/~demmel/cs267_Spr11/
6. Diaz, J., Muoz-Caro, C., & Nio, A. (2012). A survey of parallel programming models and tools in the multi and many-core era. *IEEE Transactions on Parallel & Distributed Systems, 23*(8), 1369–1386.
7. Gesbert, L., Gava, F., Loulergue, F., & Dabrowski, F. (2007). *Bulk synchronous parallel ML with exceptions* (pp. 33–42). In P. Kacsuk, T. Fahringer, Z. & Németh (Eds.), *Distributed and Parallel Systems*. New York: Springer.
8. Hudak, D. E. Introduction to the Partitioned Global Address Space (PGAS) programming model. https://www.osc.edu/sites/osc.edu/files/staff_files/dhudak/pgas-tutorial.pdf
9. IBM. Big data at the speed of business, February 2015. http://www-01.ibm.com/software/data/bigdata/. Accessed Feb 19, 2015.
10. Kessler, C., & Keller, J. (2007). Models for parallel computing: Review and perspectives. *PARS Mitteilungen, 24*, 13–29.
11. Malewicz, G., Austern, M. H., Bik, A. J. C., Dehnert, J. C., Horn, I., Leiser, N., & Czajkowski, G. (2010). Pregel: A system for large-scale graph processing. In *ACM SIGMOD10* (pp. 135–145).
12. Mishra, B. S. P., Dehuri, S., Mall, R., & Ghosh, A. (2011). Parallel single and multiple objectives genetic algorithms: A survey. *International Journal of Applied Evolutionary Computation, 2*(2), 21–58.
13. Mishra, B. S. P., & Dehuri, S. (2011). Parallel computing environments: A review. *IETE Technical Review, 28*(3), 155–162.
14. Mivule, K., Harvey, B., Cobb, C., & Sayed, H. E. (2014). A review of CUDA, mapreduce, and pthreads parallel computing models. *CoRR,* 1–10.
15. Pragmatic Programming Techniques—Atom. http://horicky.blogspot.in/2010/10/scalable-system-design-patterns.html

16. Sharma, M., & Soni, P. (2014). Comparative study of parallel programming models to compute complex algorithm. *International Journal of Computer Applications*, *96*(19), 9–12.
17. Silva, L. M., & Buyya, R. (1999). Parallel programming models and paradigms. *High Performance Cluster Computing: Architectures and Systems*, *2*, 4–27.
18. Yzelman, A. N., Bisseling, R. H., Roose, D., & Meerbergen, K. (2013). MulticoreBSP for C: A high-performance library for shared-memory parallel programming. Technical report TW 624, KU Leuven, pp. 1–15.

Chapter 3
A Deep Dive into the Hadoop World to Explore Its Various Performances

Dipayan Dev and Ripon Patgiri

Abstract Size of the data used in todays enterprises has been growing at exponential rates from last few years. Simultaneously, the need to process and analyze the large volumes of data has also increased. To handle and for analysis of large scale datasets, an open-source implementation of Apache framework, Hadoop is used now-a-days. For managing and storing of all the resources across its cluster, Hadoop possesses a distributed file system called Hadoop Distributed File System (HDFS). HDFS is written completely in Java and is depicted in such a way that in can store Big data more reliably, and can stream those at high processing time to the user applications. Hadoop has been widely used in recent days by popular organizations like Yahoo, Facebook and various online shopping market venders. On the other hand, experiments on Data-Intensive computations are going on to parallelize the processing of data. None of them could actually achieve a desirable performance. Hadoop, with its Map-Reduce parallel data processing capability can achieve these goals efficiently. This chapter initially provides an overview of the HDFS in details. The next portion of the paper evaluates Hadoops performance with various factors in different environments. The chapter shows how files less than the block size affect Hadoops R/W performance and how the time of execution of a job depends on block size and number of reducers. Chapter concludes with providing the different real challenges of Hadoop in recent days and scope for future work.

Keywords Hadoop · Big data · HDFS · Small files · Map-Reduce

D. Dev (✉) · R. Patgiri
Department of Computer Science, NIT Silchar, Silchar, India
e-mail: dev.dipayan16@gmail.com

R. Patgiri
e-mail: ripon@cse.nits.ac.in

© Springer International Publishing Switzerland 2016
B.S.P. Mishra et al. (eds.), *Techniques and Environments for Big Data Analysis*,
Studies in Big Data 17, DOI 10.1007/978-3-319-27520-8_3

3.1 Introduction

The last few years of internet technology as well as computer world has seen a lot of
growth and popularity in the field of cloud computing [9, 11]. As a consequence, the
cloud applications have given birth to Big data. Hadoop, an open source distributed
system made by Apache Software Foundation, has contributed hugely to handle
and manage such Big data [2]. Hadoop [1] has a master-slave architecture, provides
scalability and reliability to a great extent. In the last few years, this framework is
extensively used and accepted by different cloud vendors as well as Big data handler.
With the help of Hadoop, a user can deploy programs and execute processes on the
configured cluster.

The main parts of Hadoop include HDFS (Hadoop Distributed File System) [1,
13] and Map-Reduce paradigm [6]. A crucial property of Hadoop Framework is
the capacity to partition the computation and data among various nodes (known as
Data Node) and running the computations in parallel. Hadoop increases the storage
capacity, I/O BW and other computation capacity by adding commodity servers.

HDFS keeps application data and file systems meta-data [7, 23] in different
servers. Like popular DFS, viz. PVFS [4, 20], Lustre [15] and GFS [10, 16], HDFS
too saves its meta-data in a server, known as Name Node. Rest of the application
data are kept on the slaves server, known as Data Nodes.

The main purpose of the paper is to focus and reveal the various criteria that
influence the efficiency of Hadoop cluster. None of the previous papers demonstrated
this kind of work that exposes the dependencies of Hadoop efficiency.

In this chapter, we have discussed about the mechanism of HDFS and we tried
to find out the different factors on which HDFS provides maximum efficiency. The
remainder of this chapter is organized as follows. We discuss Hadoop Architecture in
Sect. 3.2. In Sect. 3.3, we discuss about file I/O operation and interaction of Hadoop
with the clients. Section 3.4 discusses three experimental works to evaluate the crucial
factors on which Hadoop clusters performance depends. The next part, Sect. 3.5 deals
with the major challenges of Hadoop field. The conclusions and future work in the
concerned issues are given in Sect. 3.6.

3.2 Related Work

Evaluation of performance in the field of Hadoop and different large-scale file systems
have been carried out many times [17, 18, 21]. HDFS performance is analyzed in
[18] whose results shows that HDFS performs poor because of the various delays
in task scheduling, fragmentation, huge disk seeks caused by disk contention under
excessive workloads. The performance of Hadoop relies heavily on the operating
system as well as on the algorithms that were employed by the disk scheduler and
various allocators.

In [21], the authors tried to integrate PVFS with Hadoop and they compared the performance with Hadoop Distributed File System using some sets of benchmarks. They paper indicates various optimizations of Hadoop that can be matched with PVFS and how durability, consistency and persistent tradeoffs made by these large-scale file system effect the cluster performance. Their results showed that, PVFS performance is as good as HDFS in Hadoop Framework.

The authors of the paper [17], analyzed the performance of BlobSeerDFS with HDFS. Their result demonstrated that, BlobSeerDFS achieved higher throughput when compared to HDFS.

In our work, we have evaluated the performance of HDFS with various parameters considering different sizes of file sizes. Compared with the conference version in [8], this chapter describes the Hadoop architecture in little more depth and extra simulation is carried out to portrait HDFS read and write behavior with different sizes of files.

3.3 Architecture of Hadoop

Hadoop framework uses pure Master/Slaves architecture (Fig. 3.1). The master nodes are given the responsibility of Name Node and Job Tracker. The main duty of Job-Trackers is to initiate tasks, track and dispatch their implementation. The charge of Data Node and Task Tracker is given to Slave nodes. The main role of TaskTracker is to process of local data and collection of all these result data as per the request received from applications and then report the performance in periodic interval to JobTracker [12]. HDFS, which seems to be heart of Hadoop, all of its charges, are given to NameNode and DataNode for fulfilling, while JobTracker and TaskTracker mainly deal with Map-Reduce application.

In this chapter, we are only dealing with HDFS architecture. So, here is a brief description about Name Node and Data Node. A short description of the client interaction with the Name Node and Data Node is also portrayed.

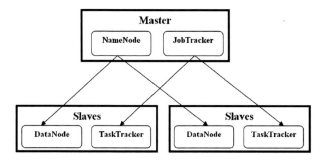

Fig. 3.1 Master-slave architecture of Hadoop cluster

3.3.1 NameNode

The HDFS namespace is a hierarchy of files and directories. Files and directories are represented on the NameNode by i-nodes, which record attributes like permissions, modification and access times, namespace and disk space quotas. The file content is split into large blocks (user defined, default 128 MB) and each block of the file is independently replicated at multiple Data Nodes (user defined, default 3). The NameNode maintains the namespace tree and the mapping of file blocks to Data Nodes. For, faster execution of the cluster operations, HDFS has to keep the whole namespace in its Main Memory. I-node data and other list of block, which belong to each file, constitute the meta-data of the name system, termed as FSImage. The un-interrupted record of the FSImage stored in the Name Nodes local filesystem is termed as checkpoint.

3.3.2 DataNodes

The block replica stored on Data Nodes is regarded as two files in the local hosts own file system. The first one constitutes the main data and second file acts as storage for meta-data of blocks. During startup each Data Node get connect to the Name Node. The phenomenon is just like a normal handshake. The purpose of this type of handshake is to verify the namespaceID and to check whether there is a mismatch between the software versions of the Data Nodes. If either does not match the same with the Name Node, that particular Data Node gets automatically shut down. A namespaceID is assigned to the file system when it is formatted each time. The namespaceID is persistently stored on all nodes of the cluster. A node with a different namespaceID will not be able to join the cluster, thus maintaining the integrity of the file system. A Data Node that is newly initialized and without any namespaceID is permitted to join the cluster and receive the clusters namespaceID. After the handshake is done, the Data Node registers with the Name Node. A storageID is allotted to the Data Node for the first time, during the registration with Name Node. Data Node periodically sends block report to the Name Node to identify the block replicas in its control. The report consists of block id, the length of each block etc. During normal operations, all the Data Nodes periodically send a response to the Name Node to confirm that, it is alive and active in operation and all the different block replicas it stores are available. If the Name Node does not receive a heartbeat from a Data Node in 10 min the Name Node considers the Data Node to be dead and the block replicas stored at that Data Node becomes inaccessible. The Name Node then schedules the things again and allocates all of those blocks to other Data Nodes, which is selected randomly.

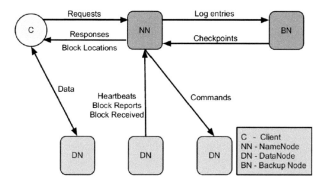

Fig. 3.2   Hadoop high-level architecture, interaction of clients with HDFS

3.3.3   HDFS Client Interaction with Hadoop

This part explains the interaction among client and Name Node and Data Node. HDFS client is the media via which a user application accesses the file system. For faster access, the namespace is saved in the Name Nodes RAM. A user references all the files and directories by paths in the namespace. It remains unknown to an application that, that file system meta-data and application data are put on separate servers as well as about the replication of blocks. During the time, an application reads a file for the first time; HDFS client gets response from the Name Node of all lists of Data Nodes that stores replicas of the blocks of that particular file. The client application checks the lists out, then a Data Node is contacted directly and requests to transfer the desired block for reading. When a client wants to write in the HDFS, it first sends a query to the Name Node to choose Data Nodes for hosting replicas of the blocks of the file. When the client application, get response from the Name Node, it start searches for that given Data Node. When if finds, the first block is filled or doesnt have enough disk space, the client requests new Data Nodes to choose for hosting replicas for the next block. Like this way, the process continues. The detail interactions among the client, the Name Node and the Data Nodes are illustrated in Fig. 3.2.

3.4   File I/O Operations and Management of Replication

3.4.1   File Read and Write

This part of paper, describes the operation of HDFS for different I/O file operations. When a client wants to write some data or add something into HDFS, he has to do the taskthrough an application. HDFS follows a single-writer, multiple-reader model [19]. When the application closes the file, the content already written cannot be manipulated or altered. However, new data can be appended in to by reopening the file All HDFS client, when tries to open a file for writing into it, is granted a lease for it; no other client can write to the file. The writing application periodically can renew the given lease by sending heartbeats to the Name Node. When the file is closed down, the lease is revoked. The lease duration is bound by a soft limit and a hard limit. The writer is granted an exclusive access for the file, as long as the soft limit persists. After the soft limits expiry comes and even then the client does not able to close the file or make renew of the lease, another client can pre-empt the lease.

The client application is also grant a hard limit of one hour. When this hard limits expiry time arrives, and here also if the client fails to renew the hard limit, HDFS taken it for granted that the client has left the network. HDFS then automatically closes that file on behalf of the writer and recovers the lease granted for it. The lease provided to a writer never does prevent any other client to read that file. HDFS follows a read by simultaneous reader at a time scheme.

An HDFS file can be defined as a collection of chunks of data or blocks. When a client application wants for a new block, Name Node does an allocation for the block, specified with a unique blockID. It then searches for a list of Data Nodes that can host the replicas of the block. Data Nodes, that act as a pipeline, generally possess has a tendency to minimize the total network distance of the last Data Node from the client.

HDFS treats all the Bytes as a sequence of packets. Bytes written by a client program is stored as a buffer. After the packet buffer is filled up (Usually 64KB), all of these are pushed Data Nodes pipeline.

When packets of bytes (data) are written into the HDFS, it never assures the reader, that he can read the file, until and unless the file is closed. There is a hush operation provided by the Hadoop, which a user application explicitly uses to see the updated file. Immediately the current packet gets pushed to the Data Nodes, and the hush operation waits until all the Data Nodes, standing in the pipeline acknowledge the successful transmission of the packet. Hadoop clusters consist of thousands of nodes. So, it is quite natural to come across failure frequently. The replicas that are stored in the Data Nodes might become corrupted as a result of memory faults, disks or several network issues. Checksums are verified by the HDFS clients to check the integrity of the data block of a HDFS file. During reading, it helps for the detection of any corrupt packet in the network. All the checksums are stored in a meta-data file in the Data-Nodes system, which is different from the blocks data file During the reading of files by HDFS, all block data and checksums are transferred to the client. It then calculates the checksums for the data and confirms that the newly calculated checksum matched with the one it received. If it doesnt match, the client informs the Name Node about the damaged replica and brings another replica from different Data Node [19].

When a file is opened by the client for reading, the client first fetches the whole list of blocks the different location of the replicas from the Name Node. The location of each block is sorted by the distance from the client. In the process of reading the content of the block, the client posses a property to read the closest replica first. If this attempt fails, it tries for the next one in sequence. If the Data Node is not available

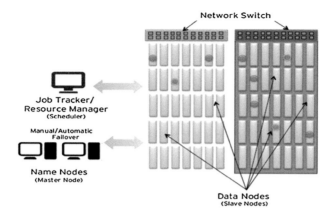

Fig. 3.3 Racks of commodity hardware and interaction with NameNode and JobTracker

or if the block replica is found to be corrupted (checksum test), the read operation then fails completely for that Data Node.

HDFS can permit a client to read the content of a file, even if it is opened for writing. But, while reading, the length of the last block, which is being written at that point of time, remains unknown to the Name Node. In this kind of case, the client asks one of the replicas for the longest length before starting to read the content.

3.4.2 Block Placement

When we want to setup a large cluster, it is never a good idea to connect all nodes to a particular switch. A better solution is that nodes of a rack should share a common switch, and switches of the rack are connected by one or more switches. The configuration of commodity hardware is done in such a way that, the network bandwidth between two nodes in the same rack is greater than network bandwidth between two nodes in different racks. Figure 3.3 shows a cluster topology, used in Hadoop architecture.

HDFS calculates the network bandwidth between two nodes by measuring the distance between them. Generally, the distance from a node to its parent node is always measured as one. A distance between two nodes can be measured by just adding up their distances to their closest node. For greater bandwidth, we should have shorter distance between two nodes. This eventually increases the capacity to transfer data.

The main concern of HDFS block replacement policy is minimization of write cost, and maximization of data reliability, scalability and increase the overall bandwidth of the cluster.

After the generation of new block, HDFS searches for a location where the writer is placed and places the first replica on that node, the 2nd and 3rd replicas are stored similarly on two different nodes in a different rack, and the rest are placed on random nodes. HDFS provides a restriction that, more than one replica cannot be stored at one node and more than two replicas cannot be stored in the same rack. The mechanism of storing the 2nd and 3rd replicas, each on different rack provides better replication management of the block replicas for a single file across the cluster.

Summing up all the above policy, the HDFS follows the following replica placement policy:

1. A single Data Node does not contain more than one replica of any block.
2. A single rack never contains more than two replicas of the same block, given then there is significant number of racks in the cluster.

3.4.3   Replication Management

The Name Node attempts to make it sure that each block of the files always has some significant number of replicas stored in the Data Nodes. Data Nodes periodically send its block report to Name Nodes. Verifying the report, the Name Node detects whether a block is under or over replicated. When Name Nodes finds it to be over replicated, it chooses a replica from any random Data Node to remove. Generally the Name Node does not prefer to reduce the number of racks which has available number of host replicas, and mainly prefers to remove from that Data Node, which has the least disk space available. The main purpose is balancing the storage utilization across all the Data Nodes, without hampering the block availability. There is a replication priority queue, which stores the blocks that is under replicated. Highest priority factor is given to the blocks, which has only one replica. On the other hand, blocks having more than two third of the specified replication factor are given the lowest priority. A thread running in background, search the replication priority queue to determine the best position for the new replicas. Block replication of HDFS also follows the same kind of policy like that of new block placement. If HDFS finds the number of existing replica of a block is one, it searches for that block and places the next replica on a rack which is different from the existing one. In case, if two replica of a block are found to be in a particular rack, the third one is kept on a different rack. Basically, the main motive here is minimizing the cost of creation of new replicas of blocks. The Name Node also does proper distribution of the block to make sure that all replicas of a particular block are not put on one single rack. If there is a situation comes, that the Name Node detects that a blocks replicas end up storing itself at one common rack, the Name Node treats it as under-replicated and eventually allocate that particular block to a different rack. When the Name Node receives the notification that a replica is created on different node, the block becomes again becomes over- replicated. So, following the previous policy written above, the Name Node, at that situation, decides to remove an old replica chosen randomly.

3.5   Performance Evaluation

In this section, the performance of Hadoop Distributed File System is evaluated in order to conclude the efficiency dependence of a Hadoop cluster.

3.5.1   Experimental Setup

For performance characterization, a 46-node Hadoop cluster was configured. The first 44 nodes provided both computation (as Map-Reduce clients) and storage resources (as Data Node servers), and the rest two nodes served as Job Tracker (Resource-Manager) and NameNode storage manager. Each node is running at 3.10 GHz clock speed and with 4 GB of RAM and a gigabit Ethernet NIC. All nodes used Hadoop framework 2.6.0, and Java 1.7.0. Ubuntu 14.04 [22] is used as out Operating System.

3.5.2   Test Using TestDFSIO to Evaluate Average I/O and Throughput of the Cluster

The test process aims at finding optimal efficiency of the performance characteristics of two different sizes of files and bottlenecks posed by the network interface. The comparison is done to check the performance between small and big files. A test of write and read between 1 GB file and 10 GB file is carried out. A total of 500 GB data is created through it. HDFS block size of 512 MB is used.

For this test we have used industry standard benchmarks: *TestDFSIO*

TestDFSIO is used to measure performance of HDFS as well as of the network and IO subsystems. The command reads and writes files in HDFS which is useful in measuring system-wide performance and exposing network bottlenecks on the NameNode and DataNodes. A majority of Map-Reduce workloads are IO bound more than compute and hence TestDFSIO can provide an accurate initial picture of such scenarios.

We executed two tests for both write and read: one for 50 files each of size 10 GB and other with 500 files each of size 1 GB.

As an example, the command for a read test may look like:

$hadoop jar Hadoop–*test*.jar TestDFSIO read nrFiles 100 fileSize 10000

This command will read 100 files, each of size 10 GB from the HDFS and there-after provides the necessary performance measures.

The command for a write test may look like:

$hadoop jar Hadoop–*test*.jar TestDFSIO write nrFiles 100 fileSize 10000

This command will write 100 files, each of size 10 GB from the HDFS and there-after provides the necessary performance measures.

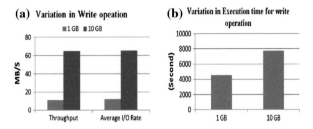

Fig. 3.4 Evaluation of write operation (**a**) the write bandwidth and throughput of (**b**) the same amount of write takes almost 10 GB file size is almost 6 times greater than 1.71 more time in case of 10 and 1 GB

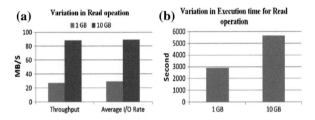

Fig. 3.5 Evaluation of read operation (**a**) reading same amount of data in 10 GB file- (**b**) the same amount of read takes almost 1.95 size offers close to 4 times more throughput and more time in case of 10 GB average bandwidth

TestDFSIO generates 1 map task per file and splits are defined such that each map gets only one file name. After every run, the command generates a log file indicating performance in terms of 4 metrics: Throughput in MBytes/s, Average IO rate in MBytes/s, IO rate standard deviation and execution time.

Output of different tests are given in Figs. 3.4 and 3.5.

To obtain a good picture of performance, each benchmark tool was run 3 times on each 1 GB file size and results were averaged to reduce error margin. The same process was carried on 10 GB file size to get data for comparison.

Experiment shows, test execution time is almost half during the 1 GB file test. This the total time it takes for the Hadoop jar command to execute.

From Figs. 3.4a and 3.5a, we can visualize that, the throughput and IO Rate too shows a significant declined in terms of both write and read for the 1 GB file test.

This is somewhat unexpected in nature. However, one major conclusion that we encountered is as follows: In these tests there is always one reducer that runs after the all map tasks have complete. The reducer is responsible for generating the result set file. It basically sums up all of these values "rate, sqrate, size, etc." from each of the map tasks. So the Throughput, IO rate, STD deviation, results are based on individual map tasks and not the overall throughput of the cluster. The nrFiles is equal to number of map tasks. In the 1 GB file test there will be $(500/6) = 83.33$ (approx) map tasks running simultaneously on each node manager node versus 8.33 map tasks on each node in the 10 GB file test. The 10 GB file test yields throughput results of 64.94 MB/s on each node manager. Therefore, the 10 GB file test yields (8.33 × 64.94 MB/s) = 540.95 MB/s per node manager. Whereas, the 1 GB file test yields throughput results of 11.13 MB/s on each nodemanager. Therefore, the 1 GB file test yields (83.33 × 11.13 MB/s) = 927.46 MB/s per node manager.

Clearly the 1 GB file test shows the maximum efficiency. However increasing the no of commodity hardware eventually decreases the execution time and increased the average IO rate. It also shows how MapReduce IO performance can vary depending on the data size, number of map/reduce tasks, and available cluster resources.

3.5.3   Dependence of Execution Time of Write Operation on No of Reducers and Block Size

A word count job is submitted and experiment is carried out on a 100 GB file, varying the no of reducers keeping with block size of HDFS fixed. The experiment carried out with 4 different types of block size, viz. 512, 256, 128 and 64 MB.

Based on the Test-Report we obtained, the charts in the Figs. 3.6, 3.7, 3.8 and 3.9 have been made and proper conclusion is followed.

Fig. 3.6 Variation of processing times with variation of reducers, keeping block size fixed at 512 MB

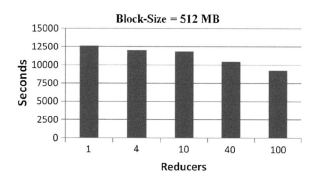

Fig. 3.7 Variation of processing times with variation of reducers, keeping block size fixed at 256 MB

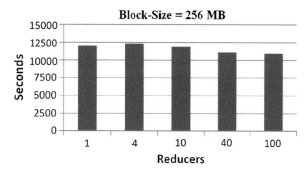

Fig. 3.8 Variation of
processing times with
variation of reducers,
keeping block size fixed
at 128 MB

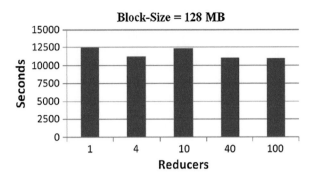

Fig. 3.9 Variation of
processing times with
variation of reducers,
keeping block size fixed
at 64 MB

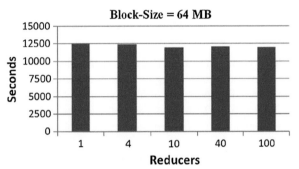

All the above graphs appear to form a uniform straight line or in some it shows a slight negative slope which indicates that with increase in number of reducer for a give block size time for processing either remains same or reduces to some extent. But, a significant negative slope is visible in Fig. 3.6, where block size equals 512 MB. On the other hand, in Figs. 3.8 and 3.9, the execution times are unpredictable and show an unexpected behavior.

It can be concluded that, for large block size, the reducers play an important role in the execution time of a Map-Reduce job. For smaller block size, the change in number of reducers doesnt bring noticeable changes in the execution time.

Moreover, for significant large files, small change in block-size doesnt lead to change the drastic change in execution time.

3.5.4 Performance Evaluation of Read and Write Operations in HDFS Varying Number of Files and Sizes

The write operation of HDFS is carried out for the different data of sizes 1, 2, 4 and 8 TB as shown in Figs. 3.10, 3.11, 3.12 and 3.13 and Tables 3.1, 3.2, 3.3 and 3.4 respectively. The block size of HDFS is kept at 64 MB for all the experiments in this subsection. In all the four figures a similar trend is observed. Figures show that,

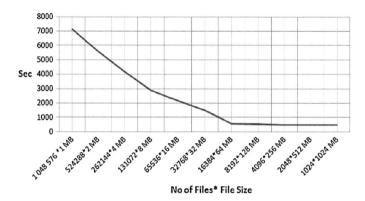

Fig. 3.10 Variation of write operation for 1 TB data size

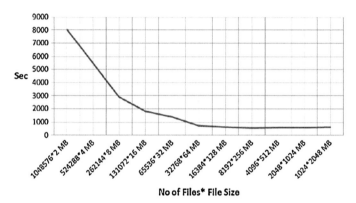

Fig. 3.11 Variation of write operation for 2 TB data size

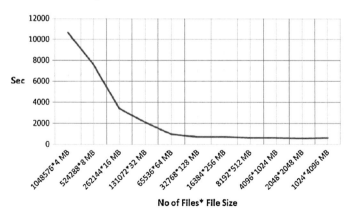

Fig. 3.12 Variation of write operation for 4 TB data size

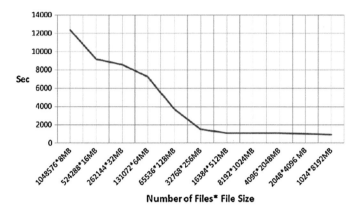

Fig. 3.13 Variation of write operation for 8 TB data size

Table 3.1 Execution time of write operation for 1 TB data size

Sl no.	No. of files	File size in MB	Execution time (in secs)
1	1048576	1	7140.68
2	524288	2	5590.058
3	262144	4	4182.967
4	131072	8	2883.34
5	65536	16	2171.219
6	32768	32	1490.87
7	16384	64	562.991
8	8192	128	544.897
9	4096	256	461.15
10	2048	512	459.18
11	1024	1024	462.07

HDFS performance is significantly poor when the file size is smaller than current block size (64 MB is our case). The execution times of the files for the write operation show a sharp decline when the size is greater than the block size.

In Figs. 3.14, 3.15, 3.16 and 3.17 and Tables 3.5, 3.6, 3.7 and 3.8, the performance of read operation for the different data of sizes 1, 2, 4 and 8 TB are shown respectively. Figure 3.14 indicates that, HDFS is taking much more time for reading 1 TB data when the file size is less than 64 MB. Whereas, when the size of the files is greater than the block size, HDFS requires much less time to read the data. Similar kind of scenario is observed in Figs. 3.15, 3.16 and 3.17.

Table 3.2 Execution time of write operation for 2 TB data size

Sl no.	No. of files	File size in MB	Execution time (in secs)
1	1048576	2	8040.32
2	524288	4	5482.437
3	262144	8	2911.926
4	131072	16	1829.159
5	65536	32	1382.462
6	32768	64	709.462
7	16384	128	623.556
8	8192	256	568.708
9	4096	512	610.539
10	2048	1024	601.506
11	1024	2048	618.796

Table 3.3 Execution time of write operation for 4 TB data size

Sl no.	No. of files	File size in MB	Execution time (in secs)
1	1048576	4	10642.859
2	524288	8	7601.318
3	262144	16	3407.575
4	131072	32	2083.141
5	65536	64	962.719
6	32768	128	723.435
7	16384	256	707.912
8	8192	512	630.882
9	4096	1024	611.868
10	2048	2048	602.868
11	1024	4096	618.948

3.6 Major Challenges in Hadoop Framework

Although the Hadoop Framework has been approved by everyone for its flexibility and faster parallel computing technology [5], there still are many problems which written in short in the following points:

1. Hadoop suffers from a irrecoverable failure called Single point of failure of Name Node. Hadoop possesses a single master server to control all the associated sub servers (slaves) for the tasks to execute, that leads to a server shortcomings like single point of failure and lacking of space capacity, which seriously affect its scalability. During the later versions of Apache Hadoop, they came out with a Secondary NameNode [14, 24] to deal with this problem.The secondary Name Node periodically check NameNodes namespace status and merges the fsimage

Table 3.4 Experimental results of write operation for 8 TB data size

Sl no.	No. of files	File size in MB	Execution time (in secs)
1	1048576	8	12360.926
2	524288	16	9161.401
3	262144	32	8592.253
4	131072	64	7265.476
5	65536	128	3684.956
6	32768	256	1514.76
7	16384	512	1092.849
8	8192	1024	1104.974
9	4096	2048	1097.671
10	2048	4096	981.7
11	1024	8192	889.511

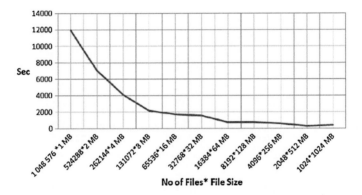

Fig. 3.14 Variation of read operation for 1 TB data size

with editlogs. It decreases the restart time of NameNode. But unfortunately is not a hot backup daemon of NameNode, not fully capable of hosting DataNodes in the absence of NameNode. So, could not resolve the SPOF of Hadoop.

2. As our experimental results show, HDFS faces huge problems, dealing with small files. HDFS data are stored in the Name Node as meta-data, and each meta-data corresponds to a block occupies about 200 Byte. Taking a replication factor of 3(default), it would take approximately 600 Byte. If there are such huge no of these kind of smaller files in the HDFS, Name Node will consume lot of space. Name Node keeps all the meta-data in its main memory, which leads to a challenging problem for the researchers [23].

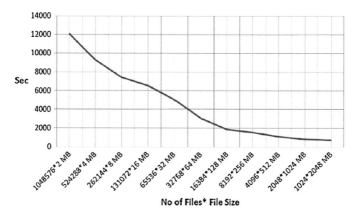

Fig. 3.15   Variation of read operation for 2 TB data size

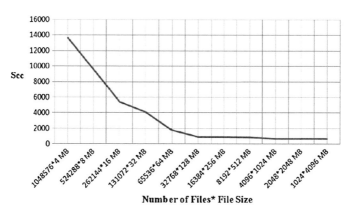

Fig. 3.16   Variation of read operation for 4 TB data size

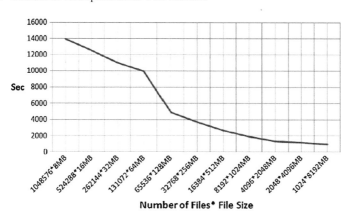

Fig. 3.17   Variation of read operation for 8 TB data size

Table 3.5 Experimental results of read operation for 1 TB data size

Sl no.	No. of files	File size in MB	Execution time (in secs)
1	1048576	1	11962.46
2	524288	2	7030.8
3	262144	4	4149.977
4	131072	8	2211.445
5	65536	16	1752.17
6	32768	32	1573.021
7	16384	64	759.939
8	8192	128	766.876
9	4096	256	611.896
10	2048	512	310.796
11	1024	1024	437.017

Table 3.6 Experimental results of read operation for 2 TB data size

Sl no.	No. of files	File size in MB	Execution time (in secs)
1	1048576	2	12062.4
2	524288	4	9288.534
3	262144	8	7452.232
4	131072	16	6535.733
5	65536	32	5041.818
6	32768	64	3048.021
7	16384	128	1864.993
8	8192	256	1542.613
9	4096	512	1099.71
10	2048	1024	812.796
11	1024	2048	747.017

3. Job Tracker at a certain time becomes extremely over loaded since it has the sole responsibility to monitor as well as dispatch simultaneously. Researchers are focusing to design a more developed version of Hadoop component for monitoring, while Job Tracker will be given the charge of overall scheduling.
4. Improving data processing performance is also a topic of major challenge for the upcoming days. A special optimization process should be assigned based on what is the actual need of application. Different experiments show that there are various scopes to increase the processing performance and thus improving time complexity of data for the execution of a particular job [3, 25].

Table 3.7 Experimental results of read operation for 4 TB data size

Sl no.	No. of files	File size in MB	Execution time (in secs)
1	1048576	4	13642.89
2	524288	8	9601.310
3	262144	16	5407.515
4	131072	32	4083.144
5	65536	64	1762.761
6	32768	128	923.675
7	16384	256	907.912
8	8192	512	830.882
9	4096	1024	711.868
10	2048	2048	702.868
11	1024	4096	718.948

Table 3.8 Experimental results of read operation for 8 TB data size

Sl no.	No. of files	File size in MB	Execution time (in secs)
1	1048576	8	13960.926
2	524288	16	12561.401
3	262144	32	10992.253
4	131072	64	9985.476
5	65536	128	4894.956
6	32768	256	3714.76
7	16384	512	2692.849
8	8192	1024	1894.974
9	4096	2048	1317.671
10	2048	4096	1181.7
11	1024	8192	989.511

3.7 Conclusion

This section of the chapter describes related future work that we are considering; Hadoop being an open source project justifies the addition of new features and changes for the sake of better scalability and file management. Hadoop is recently one of the best large-scale frameworks among managing the Big data. Still, in our experiment, we have found that, it performs poor in terms of throughput when the numbers of files are relatively larger compared to smaller numbers of files. Our experiment shows how the read/write operations of files depend on its sizes and the block size of HDFS. The performance bottlenecks are not directly imputable to application code but actually depends on numbers of data nodes available, size of files in used in HDFS and also it depends on the number of reducers used. However, the biggest issue on which we are focusing is the scalability of Hadoop Framework. The Hadoop cluster becomes unavailable when its NameNode is down.

Scalability issue of the Name Node has been a major struggle. The Name Node keeps all the namespace and block locations in its main memory. The main challenge with the Name Node has been that when its namespace table space becomes close the main memory of Name Node, it becomes unresponsive due to Java garbage collection. This scenario is bound to happen because the numbers of files used the users are increasing exponentially. Therefore, this is a burning issue in the recent days for Hadoop.

Acknowledgments The research is supported by Data Science & Analytic Lab of NIT Silchar. The authors would also like to thank the anonymous reviewers for their valuable and constructive comments on improving the chapter.

References

1. Apache Hadoop. http://hadoop.apache.org/
2. Beaver, D., Kumar, S., Li, H. C., Sobel, J., & Vajgel, P. (2010). *Finding a needle in haystack: Facebooks photo storage*. In *OSDI, ACM* (pp. 1–8).
3. Bhandarkar, M. (2010). MapReduce programming with apache Hadoop. In: *2010 IEEE International Symposium on Parallel & Distributed Processing (IPDPS)* (Vol. 1, No. 1, pp. 19–23).
4. Carns, P. H., Ligon III, W. B., Ross, R. B., & Thakur, R. (2000). PVFS: A parallel file system for Linux clusters. In *Proceedings of 4th Annual Linux Showcase and Conference* (pp. 317–327).
5. Daxin, X., & Fei, L. (2011). Research on optimization techniques for Hadoop cluster performance. *Computer Knowledge and Technology, 8*(7), 5484–5486.
6. Dean, J., & Ghemawat, S. (2004). MapReduce: Simplified data processing on large clusters. In *Proceedings Sixth Symposium Operating System Design and Implementation (OSDI 04)* (pp. 137–150).
7. Dev D., & Patgiri, R. (in press). HAR+: Archive and metadata distribution! Why not both? In *ICCCI 2015*.
8. Dev D., & Patgiri, R. (in press). Performance evaluation of HDFS in big data management. In *ICHPCA-2014*.
9. Dev, D., & Baishnab, K. L. A. (2014). Review and research towards mobile cloud computing. In *2nd IEEE International Conference on Mobile Cloud Computing, Services, and Engineering (Mobile- Cloud)* (pp. 252–256).
10. Ghemawat, S., Gobio, H. & Leung, S.-T. (2003). The google file system. In *Proceedings 19th ACM Symposium Operating Systems Principles (SOSP03)* (pp. 29–43).
11. Grobauer, B., Walloschek, T., & Stocker, E. Understanding cloud computing vulnerabilities. In *IEEE International Conference on Security & Privacy* (vol. 9, pp. 50–57).
12. Guilan, X., & Shengxian, L. (2010). Research on applications based on Hadoop MapReduce model. *Microcomputer & Its Applications (8)*, 4–7.
13. Hadoop Distributed File System Rebalancing Blocks. (2012). http://developer.yahoo.com/hadoop/tutorial/module2.html#rebalancing.
14. HDFS Federation. (2012). http://hadoop.apache.org/common/docs/r0.23.0/hadoop-yarn/hadoop-yarn-site/Federation.html.
15. Lustre File System. http://www.lustre.org.
16. McKusick, K., & Quinlan, S. G. F. S. (2010). Evolution on Fast-Forward. *Communication of the ACM, 53*(3), 42–49.

17. Nicolae, B., Moise, D., Antoniu, G., Boug, L., & Dorier, M. (2010). BlobSeer: Bringing high throughput under heavy concurrency to Hadoop Map/Reduce applications. In *Proceedongs 24th IEEE Interational Parallel and Distributed Processing Symposium (IPDPS 2010)*.
18. Shafer, J. A. (2010). *Storage architecture for data-intensive computing*. Ph.D. thesis, Rice University. Advisor-Rixner, Scott.
19. Shvachko, K., Kuang, H., Radia, S., & Chansler, R. (2010). The Hadoop distributed file system. In *IEEE 26th Symposium on Mass Storage Systems and Technologies (MSST)* (Vol. 1, No. 10, pp. 3–7).
20. Tantisiriroj, W., Patil, S., & Gibson, G. (2008, October). Data-intensive file systems for Internet services: A rose by any other name. Technical Report CMUPDL- 08–114, Parallel Data Laboratory, Carnegie Mellon University, Pittsburgh, PA.
21. Tantisiriroj, W., Patil, S., Gibson, G., Son, S. W., Lang, S. J., & Ross, R. B. On the duality of data-intensive file system design: Reconciling HDFS and PVFS. In *SC11*.
22. Ubuntu. http://releases.ubuntu.com/14.04/.
23. Weil, S. A., Pollack, K. T., Brandt, S. A., & Miller, E. L. (2004). Dynamic metadata management for petabyte-scale file systems. In *ACM/IEEE SC* (pp. 4–12).
24. White, T. (2009). Hadoop, guide, The Definitive, & Inc, O' Reilly Media. (1005). Gravenstein Highway North, Sebastopol. CA, 95472.
25. Yan, J., Yang, X., Gu, R., Yuan, C., & Huang, Y. (2012). Performance optimization for short MapReduce job execution in Hadoop. In: *2012 Second International Conference on Cloud and Green Computing (CGC)* (Vol. 688, No. 694, pp. 1–3).

Chapter 4
Natural Language Processing and Machine Learning for Big Data

Joy Mustafi

Abstract This chapter is focused on how Big data challenges can be handled from the data science perspective. The data available for analysis are in different forms in terms of volume, velocity, variety, and veracity. The objective is to resolve some of these real world problems using natural language processing, where the unstructured data can be transformed into meaningful structured information; and machine learning to get more insights out of the information available or derived. The combination of multiple algorithms can actually play a major role in the overall field of cognitive computing. The chapter fairly covers important methodologies where, what and when to apply. Some open research problems are shared for the budding data scientists. This chapter may be referred as the basic introduction to data science.

4.1 Introduction to NLP and ML

On February 14, 2011, IBM Watson changed history introducing a system that rivaled a human's ability to answer questions posed in natural language with speed, accuracy and confidence by transforming how organizations think, act, and operate; learning through interactions and delivering evidence based responses driving better outcomes [1] (Fig. 4.1).

The data available for analysis are in different forms in terms of volume, velocity, variety and veracity. Which is also referred as Big data! There are many technologies for handling these enormous number of structured and unstructured data available from various source—starting from social network server to mobile micro memory card [4]. Therefore to handle with the unstructured data or information in natural language it becomes a challenge. Data scientists across the globe are working on the algorithms on natural language processing and machine learning to retrieve the relevant information from the textual data, which is important (Fig. 4.2).

J. Mustafi (✉)
Cognitive Computing—Data Science—Advanced Analytics, Bangalore, India
e-mail: jmustafi@in.ibm.com

© Springer International Publishing Switzerland 2016   53
B.S.P. Mishra et al. (eds.), *Techniques and Environments for Big Data Analysis*,
Studies in Big Data 17, DOI 10.1007/978-3-319-27520-8_4

Fig. 4.1 IBM Watson introducing Big data

Fig. 4.2 Various sources of data

4.1.1   Overview of Natural Language Processing (NLP)

Natural language processing (NLP) is a field of computer science, artificial intelligence, and linguistics concerned with the interactions between computers and human languages [2]. As such, NLP is related to the area of human–machine interaction. Many challenges in NLP involve natural language understanding, that is, enabling

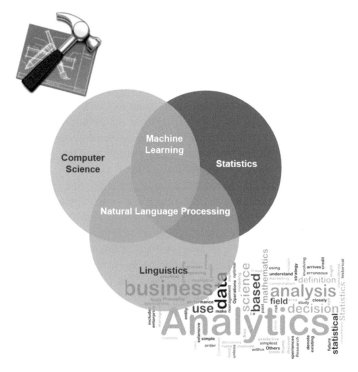

Fig. 4.3 The domain of natural language processing

computers to derive meaning from natural language input, and others involve natural language generation by computer.

Modern NLP algorithms are based on machine learning, especially statistical machine learning. The traditional NLP algorithms are linguistic-based or rule-based. There are hybrid models as well—a combination of statistical and rule-based systems. The paradigm of machine learning is different from that of most prior attempts at language processing. Prior implementations of language-processing tasks typically involved the direct hand coding of large sets of rules. The machine-learning paradigm calls instead for using general learning algorithms—often, although not always, grounded in statistical inference—to automatically learn such rules through the analysis of large corpora of typical real-world examples. A corpus (plural: corpora) is a set of documents (or sometimes, individual sentences) that have been hand-annotated with the correct values to be learned (Fig. 4.3).

4.1.2   Background of Machine Learning (ML)

Machine learning is a scientific discipline that explores the construction and study of algorithms that can learn from available information. Such algorithms operate by building a model based on inputs and using that to make predictions or decisions, rather than following only explicitly programmed instructions. ML can be considered a subfield of computer science and statistics. It has strong ties to artificial intelligence and optimization, which deliver methods, theory and application domains to the field. Machine learning is employed in a range of computing tasks where designing and programming explicit, rule-based algorithms is infeasible.

Automatic learning procedures can make use of statistical inference algorithms to produce models that are robust to unfamiliar input (e.g. containing words or structures that have not been seen before) and to erroneous input (e.g. with misspelled words or words accidentally omitted). Generally, handling such input gracefully with hand-written rules—or more generally, creating systems of hand-written rules that make soft decisions—is extremely difficult, error-prone and time-consuming.

Systems based on automatically learning the rules can be made more accurate simply by supplying more input data. However, systems based on hand-written rules can only be made more accurate by increasing the complexity of the rules, which is a much more difficult task. In particular, there is a limit to the complexity of systems based on hand-crafted rules, beyond which the systems become more and more unmanageable. However, creating more data to input to machine-learning systems simply requires a corresponding increase in the number of man-hours worked, generally without significant increases in the complexity of the annotation process.

4.2   Unstructured Data Analysis Using NLP

Variety—The next aspect of Big data is its variety. This means that the category to which Big data belongs to is also a very essential fact that needs to be known by the data analysts. This helps the people, who are closely analyzing the data and are associated with it, to effectively use the data to their advantage and thus upholding the importance of the Big data [6].

Unstructured data (or unstructured information) refers to information that either does not have a pre-defined data model or is not organized in a pre-defined manner. Unstructured information is typically text-heavy, but may contain data such as dates, numbers, and facts as well. This results in irregularities and ambiguities that make it difficult to understand using traditional computer programs as compared to data stored in fielded form in databases or annotated (semantically tagged) in documents.

4.2.1   NLP Majors Tasks and Features

Natural Language Understanding (NLU): Converting chunks of text into more formal representations such as first-order logic structures that are easier for computer programs to manipulate. Natural language understanding involves the identification of the intended semantic from the multiple possible semantics which can be derived from a natural language expression which usually takes the form of organized notations of natural languages concepts.

Natural Language Generation (NLG): Convert information from computer databases into readable human language.

Morphological Analysis and Segmentation: Separate words into individual morphemes and identify the class of the morphemes. The difficulty of this task depends greatly on the complexity of the structure of words of the language being considered. Some language like English has fairly simple morphology, especially inflectional morphology, and thus it is often possible to ignore this task entirely and simply model all possible forms of a word (e.g. eat, eats, ate, eaten, eating) as separate words.

Part-of-Speech (POS) Tagging: Given a sentence, determine the part of speech for each word. Many words, especially common ones, can serve as multiple parts of speech. For example, book can be a noun (the book on the table) or verb (to book a flight); set can be a noun, verb or adjective; and out can be any of at least five different parts of speech [7].

Parsing: Determine the parse tree (grammatical analysis) of a given sentence. The grammar for natural languages is ambiguous and typical sentences have multiple possible analyses. In fact, perhaps surprisingly, for a typical sentence there may be thousands of potential parses (most of which will seem completely nonsensical to a human) [8].

Coreference Resolution: Given a sentence or larger chunk of text, determine which words (mentions) refer to the same objects (entities). Anaphora resolution is a specific example of this task, and is specifically concerned with matching up pronouns with the nouns or names that they refer to. The more general task of co-reference resolution also includes identifying so-called bridging relationships involving referring expressions.

Discourse Analysis: This id for identifying the discourse structure of connected text, i.e. the nature of the discourse relationships between sentences (e.g. elaboration, explanation, contrast). Another possible task is recognizing and classifying the speech acts in a chunk of text (e.g. yes-no question, content question, statement, assertion, etc.).

Word Sense Disambiguation: Many words have more than one meaning; we have to select the meaning which makes the most sense in context. For this problem, we are typically given a list of words and associated word senses.

Named Entity Recognition (NER): Given a stream of text, determine which items in the text map to proper names, such as people or places, and what the type of each such name is (e.g. person, location, organization). Note that, although capitalization can aid in recognizing named entities in languages such as English, this information cannot aid in determining the type of named entity, and in any case is often inaccurate or insufficient.

Automatic Summarization: Produce a readable summary of a chunk of text. Which is used to provide summaries of text of a known type, such as articles in the financial section of a newspaper.

Sentiment Analysis: Extract subjective information usually from a set of documents, often using online reviews to determine polarity about specific objects. It is especially useful for identifying trends of public opinion in the social media, for the purpose of marketing.

Machine Translation (MT): Automatically translate text from one human language to another. This is one of the most difficult problems, and is a member of a class of problems colloquially termed AI-complete, i.e. requiring all of the different types of knowledge that humans possess (grammar, semantics, facts about the real world, etc.) in order to solve properly [3].

Question Answering: Given a human-language question, determine its answer. Typical questions have a specific right answer (such as 'What is the capital of India?'). e.g. IBM Watson.

4.2.2 Morphological Analysis

In linguistics, morphology is the identification, analysis and description of the structure of a given language's morphemes and other linguistic units, such as root words, affixes, parts of speech, intonation/stress, or implied context (words in a lexicon are the subject matter of lexicology) (Fig. 4.4).

Analysis consists of characteristic properties (**word formation**) like root or stem, lexical category, gender, number, person… etc.

Example: eaten
    <Root> → eat
    <Lexical Category> → Verb

- **Inflectional Morphology**
  The root has same lexical category as the word-form

Example: eat → Verb, eating → Verb

- **Derivational Morphology**
  Lexical category changes

Example: eat → Verb, eater → Noun

eat | eats | ate | eaten | eating → eat

Fig. 4.4 Morphology

4.2.3   POS Tagging

In corpus linguistics, Part-Of-Speech Tagging (POS tagging or POST), also called grammatical tagging, is the process of marking up a word in a text (corpus) as corresponding to a particular part of speech, based on both its definition, as well as its context—i.e. relationship with adjacent and related words in a phrase, sentence, or paragraph (Figs. 4.5 and 4.6).

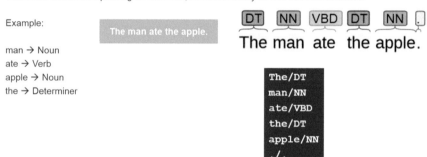

Part of Speech (POS) Tags in English:
Broadly classified as noun, verb, adjective and adverb.

Words are classified depending on their **role**, both individually as well as in the sentence.

Example:

The man ate the apple.

man → Noun
ate → Verb
apple → Noun
the → Determiner

The man ate the apple.
DT NN VBD DT NN .

The/DT
man/NN
ate/VBD
the/DT
apple/NN
. / .

Fig. 4.5 POS tagging of words

CC	Coordinating conjunction	PRP$	Possessive pronoun
CD	Cardinal number	RB	Adverb
DT	Determiner	RBR	Adverb, comparative
EX	Existential there	RBS	Adverb, superlative
FW	Foreign word	RP	Particle
IN	Preposition or subordinating conjunction	SYM	Symbol
JJ	Adjective	TO	to
JJR	Adjective, comparative	UH	Interjection
JJS	Adjective, superlative	VB	Verb, base form
LS	List item marker	VBD	Verb, past tense
MD	Modal	VBG	Verb, gerund or present participle
NN	Noun, singular or mass	VBN	Verb, past participle
NNS	Noun, plural	VBP	Verb, non-3rd person singular present
NNP	Proper noun, singular	VBZ	Verb, 3rd person singular present
NNPS	Proper noun, plural	WDT	Wh-determiner
PDT	Predeterminer	WP	Wh-pronoun
POS	Possessive ending	WP$	Possessive wh-pronoun
PRP	Personal pronoun	WRB	Wh-adverb

Fig. 4.6 Abbreviations for POS tags

A natural language parser is a program that works out the **grammatical structure** of sentences, for instance, which groups of words go together (as "phrases") and which words are the **subject** or **object** of a verb.

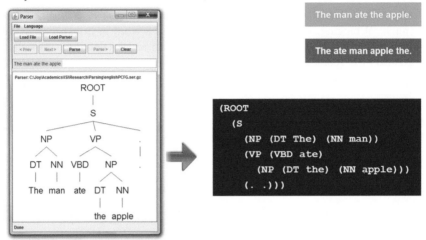

Fig. 4.7 The process of parsing

4.2.4   Parsing

Parsing/Syntactic Analysis: Parsing or Syntactic Analysis is the process of analyzing a string of symbols, either in natural language or in computer languages, according to the rules of a formal grammar (Fig. 4.7).

4.2.5   Dependency Analysis

A representation of grammatical relations between words in a sentence. They have been designed to be easily understood and effectively used by people who want to extract textual relations. In general, dependencies are triplets: Name of the relation, governor and dependent (Fig. 4.8).

4.2.6   Semantics and Word Sense Disambiguation

In computational linguistics, word-sense disambiguation (WSD) is an open problem of natural language processing and ontology, which governs the process of identifying which sense of a word (i.e. meaning) is used in a sentence, when the word has multiple meanings. The solution to this problem impacts other computer-related writing, such

Dependency analysis using the parse tree was used to determine the **relationships** between different words and phrases had within a sentence.

With target occurrences occurring in noun phrases, words were only considered as relating to the target if they satisfied at least one of the following:

1. The word is contained **within a noun phrase** where the target occurs.
2. The word is contained **within a phrase with the same parent node as a noun phrase** where the target occurs, and the phrase occurs after the target in the text.

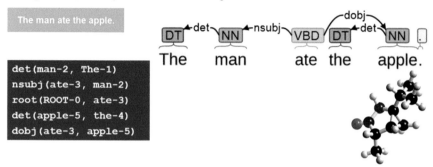

Fig. 4.8 Dependency analysis among words

as discourse, improving relevance of search engines, anaphora resolution, coherence, inference et cetera.

The human brain is remarkably good at word-sense disambiguation. The fact that natural language is formed in a way that requires so much of it is a reflection of that neurologic reality. In other words, human language developed in a way that reflects (and also has helped to shape) the innate ability provided by the brain's neural networks. In computer science and the information technology that it enables, it has been a long-term challenge to develop the ability in computers to do natural language processing and machine learning.

Research has progressed steadily to a point where WSD systems achieve sufficiently high levels of accuracy on a variety of word types and ambiguities. A rich variety of techniques have been researched, from dictionary-based methods that use the knowledge encoded in lexical resources, to supervised machine learning methods in which a classifier is trained for each distinct word on a corpus of manually sense-annotated examples, to completely unsupervised methods that cluster occurrences of words, thereby inducing word senses. Among these, supervised learning approaches have been the most successful algorithms to date (Figs. 4.9 and 4.10).

In linguistics, semantic analysis is the process of relating syntactic structures, from the levels of phrases, clauses, sentences and paragraphs to the level of the writing as a whole, to their language-independent **meanings**.

The task is to determine which of the **various senses** of a word are **involved in context**.

- **Disambiguating using POS Tags**
  Example: play as verb is to play, as noun is drama

- **Collocation**
  Example: handsome person, handsome salary operation (in different domain)

- **Selection Restriction**
  Example: I hate washing dishes, I love spicy dishes

The man ate the apple.   The apple ate the man.

Fig. 4.9 Semantic analysis

Fig. 4.10 Information—the ultimate goal   **The Bottom Line**

4.3 Information Retrieval Using Machine Learning

Many different classes of machine learning algorithms have been applied to NLP tasks. These algorithms take as input a large set of features that are generated from the input data. Some of the earliest-used algorithms, such as decision trees, produced systems of hard if-then rules similar to the systems of hand-written rules that were then common. Increasingly, however, research has focused on statistical models, which make soft, probabilistic decisions based on attaching real-valued weights to each input feature. Such models have the advantage that they can express the relative certainty of many different possible answers rather than only one, producing more reliable results when such a model is included as a component of a larger system.

The learning procedures used during machine learning automatically focus on the most common cases, whereas when writing rules by hand it is often not obvious at all where the effort should be directed.

Example applications include spam filtering, optical character recognition (OCR), search engines and computer vision. Machine learning is sometimes conflated with data mining, although that focuses more on exploratory data analysis.

Information retrieval (IR): This is concerned with storing, searching and retrieving information. It is a separate field within computer science (closer to databases), but IR relies on some NLP methods (for example, stemming).

Information Extraction (IE): This is concerned in general with the extraction of semantic information from text. This covers tasks such as named entity recognition, Coreference resolution, relationship extraction, etc.

4.3.1   Machine Learning Approaches

Decision Tree Learning: Decision tree learning uses a decision tree as a predictive model, which maps observations about an item to conclusions about the item's target value.

Association Rule Learning: Association rule learning is a method for discovering interesting relations between variables in large databases.

Artificial Neural Network: An artificial neural network (ANN) learning algorithm, usually called neural network (NN), is a learning algorithm that is inspired by the structure and functional aspects of biological neural networks. Computations are structured in terms of an interconnected group of artificial neurons, processing information using a connectionist approach to computation. Modern neural networks are non-linear statistical data modeling tools. They are usually used to model complex relationships between inputs and outputs, to find patterns in data, or to capture the statistical structure in an unknown joint probability distribution between observed variables.

Inductive Logic Programming: Inductive logic programming (ILP) is an approach to rule learning using logic programming as a uniform representation for input examples, background knowledge, and hypotheses. Given an encoding of the known background knowledge and a set of examples represented as a logical database of facts, an ILP system will derive a hypothesized logic program that entails all positive and no negative examples. Inductive programming is a related field that considers any kind of programming languages for representing hypotheses (and not only logic programming), such as functional programs.

Support Vector Machines: Support vector machines (SVMs) are a set of related supervised learning methods used for classification and regression. Given a set of training examples, each marked as belonging to one of two categories, an SVM training algorithm builds a model that predicts whether a new example falls into one category or the other.

Cluster Analysis: Cluster analysis is the assignment of a set of observations into subsets (called clusters) so that observations within the same cluster are similar, while observations drawn from different clusters are dissimilar. Different clustering techniques make different assumptions on the structure of the data, often defined by some similarity metric and evaluated for example by internal compactness (similarity between members of the same cluster) and separation between different clusters. Other methods are based on estimated density and graph connectivity. Clustering is a method of unsupervised learning, and a common technique for statistical data analysis.

Bayesian Network: A Bayesian network, belief network or directed acyclic graphical model is a probabilistic graphical model that represents a set of random variables and their conditional independencies via a directed acyclic graph (DAG). For example, a Bayesian network could represent the probabilistic relationships between diseases and symptoms. Given symptoms, the network can be used to compute the probabilities of the presence of various diseases. Efficient algorithms exist that perform inference and learning.

Reinforcement Learning: Reinforcement learning is concerned with how an agent ought to take actions in an environment so as to maximize some notion of long-term reward. Reinforcement learning algorithms attempt to find a policy that maps states of the world to the actions the agent ought to take in those states. Reinforcement learning differs from the supervised learning problem in that correct input/output pairs are never presented, nor sub-optimal actions explicitly corrected.

Representation Learning: Several learning algorithms, mostly unsupervised learning algorithms, aim at discovering better representations of the inputs provided during training. Classical examples include principal components analysis and cluster analysis. Representation learning algorithms often attempt to preserve the information in their input but transform it in a way that makes it useful, often as a pre-processing step before performing classification or predictions, allowing to reconstruct the inputs coming from the unknown data generating distribution, while not being necessarily faithful for configurations that are implausible under that distribution.

Similarity Learning: In this problem, the learning machine is given pairs of examples that are considered similar and pairs of less similar objects. It then needs to learn a similarity function (or a distance metric function) that can predict if new objects are similar. It is sometimes used in Recommendation systems.

Genetic Algorithms: A genetic algorithm (GA) is a search heuristic that mimics the process of natural selection, and uses methods such as mutation and crossover to generate new genotype in the hope of finding good solutions to a given problem. In machine learning, genetic algorithms found some uses in the 1980s and 1990s.

4.3.2   Advanced Analytics Using Machine Learning

Analytics are usually based on modeling requiring extensive computation and thus the algorithms and software used for analytics tend to bridge the disciplines of computer science, statistics, and mathematics.

A common application of analytics include the study of business data with an eye to predicting and improving business performance in the future.

Business analytics can answer questions like why is this happening, what if these trends continue, what will happen next (that is, predict), what is the best that can happen (that is, optimize) etc.

Analytics is the discovery and communication of meaningful patterns in data. Advanced analytics refers to the skills, technologies, applications and practices for continuous iterative exploration and investigation of past events to gain insight and drive planning. Business analytics focuses on developing new insights and understanding of business performance based on data and statistical methods. Predictive analytics makes extensive use of data, statistical and quantitative analysis, explanatory and predictive modeling, and fact-based management to drive decision making.

4.3.3 Model Building

A model is a proposed abstraction of reality. It represents the principles with essential characteristics of behavior or phenomenon in the real world in a simplified way. Though the complex mental processes are not easy to model quantitatively, higher level of abstraction is necessary for the construction of conceptual models. Models are more precise than verbal descriptions and offer greater manipulability. Steps:

- Selection of Topic
- Objective, Hypothesis, Research Questions
- Construct Variables
- Sampling, Data Collection, Master Sheet
- Statistical Analysis, Mathematical Modeling
- Interpretation, Report Writing

Model building leading to hypothesis specification is done at an early stage of research (conceptual part of research, theory building).

To represent the reality: To what extent observed results depict the reality (empirical part of research, theory testing).

Actual knowledge about reality exists outside. The researcher formulates beliefs about that reality. The belief statements about the happenings of reality are the basis of conjectures/hypotheses/research questions. Then these are tested collecting data, analyzing data, and reporting results. If the results supports the beliefs, knowledge generated is accepted.

Multiple Regression: A single metric dependent variable is predicted by several metric independent variables (Fig. 4.11).

Discriminant Analysis: A single, non-metric (categorical) dependent variable is predicted by several metric independent variables.

Logistic Regression: A single non-metric dependent variable is predicted by several metric independent variables. This technique is similar to Discriminant Analysis, but relies on calculations more like regression.

Multivariate Analysis of Variance (MANOVA): Several metric dependent variables are predicted by a set of non-metric (categorical) independent variables.

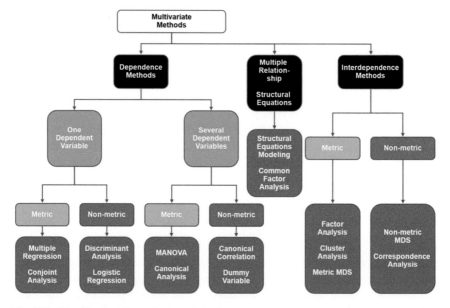

Fig. 4.11 Classification of multivariate methods

Canonical Analysis: Several metric dependent variables are predicted by several metric independent variables.

Conjoint Analysis: Used to understand respondents preferences for products and services.

Structural Equations Modeling (SEM): Estimates multiple, interrelated dependence relationships based on two components: Measurement Model and Structural Model.

Exploratory Factor Analysis: Analyzes the structure of the interrelationships among a large number of variables to determine a set of common underlying dimensions (factors).

Cluster Analysis: Groups objects (respondents, products, firms, variables, etc.), so that each object is similar to the other objects in the cluster and different from objects in all the other clusters.

Multidimensional Scaling (MDS): Identifies unrecognized dimensions that affect purchase behavior based on customer judgments of: Similarities or Preferences; and transforms these into distances represented as perceptual maps.

Correspondence Analysis: Uses non-metric data and evaluates either linear relationships or non-linear relationships in an effort to develop a perceptual map representing the association between objects (firms, products, etc.) and a set of descriptive characteristics of the objects.

4.4   Statistical Packages for Data and Text Mining

Data mining an interdisciplinary subfield of computer science, is the computational process of discovering patterns in large data sets involving methods at the intersection of artificial intelligence, machine learning, statistics, and database systems. The overall goal of the data mining process is to extract information from a data set and transform it into an understandable structure for further use. Aside from the raw analysis step, it involves database and data management aspects, data preprocessing, model and inference considerations, interestingness metrics, complexity considerations, post-processing of discovered structures, visualization, and online updating.

4.4.1   Data Mining Packages

An industry standard modeling software typically have the features like:

- Boost profits and reduce costs by targeting only the most valuable customers.
- Forecast future trends to better plan organizational strategies, logistics, and manufacturing processes.
- Detect fraud and minimize business risk.
- Analyze outcomes, such as patient survival rates or good/bad credit risks.
- Report results clearly and efficiently.
- Understand which characteristics consumers relate most closely to their brand.
- Identify groups, discover relationships between groups, and predict future events.
- Access, organize, and model all types of data from within a single intuitive visual interface. Build reliable models and deploy results quickly to meet business goals.
- Easy-to-use graphical interface puts the power of data mining in the hands of business users to discover new insight and increase productivity.
- Optimization techniques for large datasets, including boosting and bagging—that improve model stability and accuracy. Visualization for key algorithms, including Neural Net and Decision Tree etc.
- Industry-specific pre-build text analysis packages for advertising, banking, insurance and others.
- Improved performance for large datasets. Text/Entity Analytics and Social Network Analysis (Fig. 4.12).

4.4.2   Text Mining Programs

Text Mining refers to extracting usable knowledge from unstructured text data, through identification of core concepts, opinions and trends, to drive better business

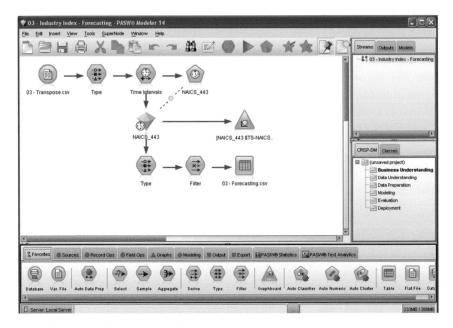

Fig. 4.12 A sample data mining and forecasting software

decisions across the enterprise. Text mining, also referred to as text data mining, roughly equivalent to text analytics, refers to the process of deriving high-quality information from text. High-quality information is typically derived through the devising of patterns and trends through means such as statistical pattern learning. Text mining usually involves the process of structuring the input text (usually parsing, along with the addition of some derived linguistic features and the removal of others, and subsequent insertion into a database), deriving patterns within the structured data, and finally evaluation and interpretation of the output. 'High quality' in text mining usually refers to some combination of relevance, novelty, and interestingness. Typical text mining tasks include text categorization, text clustering, concept/entity extraction, production of granular taxonomies, sentiment analysis, document summarization, and entity relation modeling (i.e., learning relations between named entities).

A typical application is to scan a set of documents written in a natural language and either model the document set for predictive classification purposes or populate a database or search index with the information extracted (Fig. 4.13).

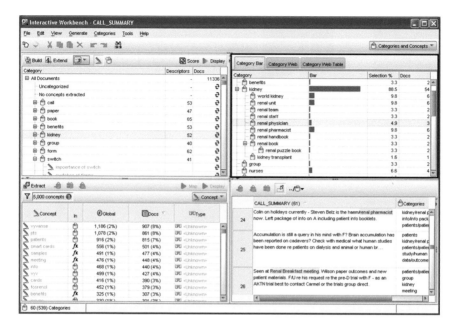

Fig. 4.13 A sample text-mining application

4.5 Research Problems in Big Data Analytics

There are various open research open problems on Big data analytics. Since, social network is playing a major role in this niche area, and lot of insights can be drawn over wide spectrum of data, some examples are described in this field.

4.5.1 Social Influencers Detection Within an Organization

In any organization there are social influencers, the informal leaders with the ability to promote change and encourage action by others without the support of the formal hierarchy. Since influencers can come from all over an organization, and usually don't correlate to a specific title or role, most of them are hidden (Fig. 4.14).

A system to identify hidden gems of the organization—social influencers using various data sources such as GPS data received from smart phones of employees, communication between employees in the form of chats and emails and data from enterprise social networking site. System uses these various data sources to compensate in a situation where one or other data sources are missing. From these data sources mobile based temporal features (start time, end time, time duration etc.), mobile based spatial features (location name, type of location, proximity based information etc.), text based communication features (advise related, informal talk related,

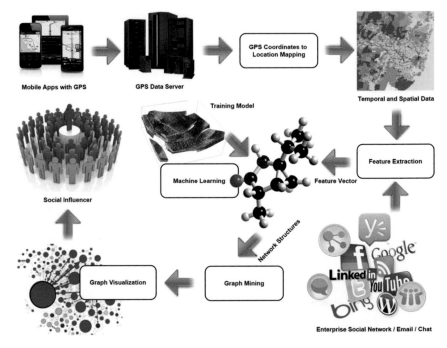

Fig. 4.14 Representation of social influencers

opinion related etc.) and social networking sites based feature (no. of recommendations, no. of likes, no. of followers etc.) are extracted. By applying Machine Learning using these features, various network structures are identified and then graph mining algorithms [5] like page rank is applied to identify social influencers.

4.5.2 Social Media Marketing Using Intention Analysis

Social Media Marketers have been using push marketing strategy in which ads are targeted to the user based on user's profile and interests that are obtained from actions such as likes, activities in groups etc. or sentiment of his/her opinions. Since this approach is focusing on demand generation, and not on capturing demand latent in text, generally the conversion rate and hence ROI is very low (Fig. 4.15).

A system that continuously monitor user generated text of social media sites obtained via data provider and extract intentions from the text using Natural Language Processing and Machine Learning. Based on the type of intent such as purchase, wish, plan etc., type of event involved in intent such as travel, buy, meet etc., types of entities involved in intent such as product, service, place etc. and other information such as location of user, interest graph of user etc., system will select or generate relevant, personalized message on behalf of marketers automatically and send to the

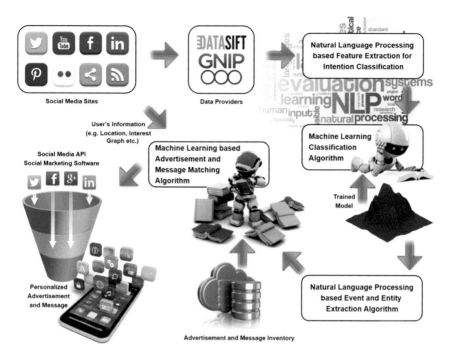

Fig. 4.15 The process of social media marketing

user instantly. This system can be used in various phases of marketing such as lead generation, social media advertisement, social customer relationship management etc. Since in this approach we are capturing user's immediate or future demand (pull marketing strategy), if combined with or replacing the push marketing strategy, then there is a promise of relatively high conversion rate and hence ROI.

4.5.3 Cognitive Computing for Education

The application will not be automatically generating content and pose questions. Only the users will be posing the questions and the app will be answering those. Adding filters based on content is going to be a new feature and can be added as a post processor to ingestion. Aggregation of content (text summarization) is a very hard problem to solve, hence the answers would be filtered based on category but will not be modified. Multi-lingual and multi-modal should be kept for future vision. Internet is generally not allowed for kids. Also, mobiles are not allowed generally in schools. So, the app should be a desktop or web version on intranet. Focus should not be content that will be ingested however we can concentrate on some test data for the designed models. The more focus should be writing filters for any type of content that

Fig. 4.16 Cognitive computing

can be extended for different kind of ingested content. Sign up flow questions need not be related to hobbies/interests because we are not looking at segregating content at App based on users' interests. Rather the idea behind building up the profile is to assess the IQ level of the user. This will help in writing out the filtering rules. Salient Features (Fig. 4.16):

- the users based on their demography will be categorized and similarly the content will also be segregated during ingestion for personalization
- answering stateless questions based on the user category
- desktop client application or web based interface deployable on server
- supported language—English only
- completely preloaded with textual information with absolutely no access to internet.

Appendix

- Carrot2: Text and search results clustering framework
- ELKI: A university research project with advanced cluster analysis and outlier detection methods
- GATE: A natural language processing and language engineering tool
- KNIME: A user friendly and comprehensive data analytics framework
- MLPACK library: A collection of ready-to-use machine learning algorithms (Fig. 4.17)

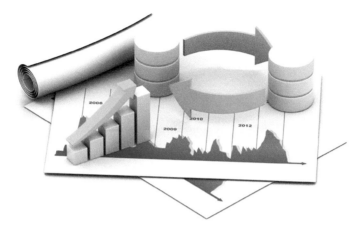

Fig. 4.17 Graphical view of databases

- Massive Online Analysis (MOA): A real-time Big data stream mining with concept drift tool
- NLTK (Natural Language Toolkit): A suite of libraries and programs for symbolic and statistical NLP
- OpenNN: Open neural networks library
- Orange: A component-based data mining and machine learning software
- R: A programming language and software environment for statistical computing and data mining
- RapidMiner: an environment for machine learning and data mining experiments
- UIMA (Unstructured Information Management Architecture): A component framework for analyzing unstructured content such as text, audio and video—originally developed by IBM
- Weka: A suite of machine learning software
- Angoss KnowledgeSTUDIO: Data mining tool provided by Angoss
- Clarabridge: Enterprise class text analytics solution
- HP Vertica Analytics Platform: Data mining software provided by HP
- IBM SPSS Modeler: Data mining software provided by IBM
- KXEN Modeler: Data mining tool provided by KXEN
- Grapheme: Data mining and visualization software provided by iChrome
- LIONsolver: An integrated software for data mining, business intelligence, and modeling
- Microsoft Analysis Services: Data mining software provided by Microsoft
- NetOwl: Suite of multilingual text and entity analytics products that enable data mining
- Oracle Data Mining: Data mining software by Oracle
- SAS Enterprise Miner: Data mining software provided by the SAS Institute
- STATISTICA Data Miner: Data mining software provided by StatSoft
- Qlucore Omics Explorer: Data mining software provided by Qlucore.

Disclaimer: The views expressed herein are of the author and do not necessarily represent the views of his employer.

References

1. IBM Watson. http://www.ibm.com/smarterplanet/us/en/ibmwatson/.
2. Jurafsky, D., & Martin, J. H. Speech and language processing speech and language processing.
3. Mustafi, J., & Chaudhuri, B. B. (2008). A proposal for standardization of English to Bangla transliteration and Bangla universal editor. *Lang. India*, *8*, 5. (May 2008).
4. Mustafi, J., Mukherjee, S., & Chaudhuri, A. (2002). Grid computing: The future of distributed computing for high performance scientific and business applications. In *Lecture Notes in Computer Science, International Workshop on Distributed Computing*, Springer, vol. 2571, pp. 339–342.
5. Mustafi, J., Parikh, A., Polisetty, A., Agarwalla, L., & Mungi, A. (2014). Thinkminers: Disorder recognition using conditional random fields and distributional semantics. In *The 8th International Workshop on Semantic Evaluation (SemEval—COLING)*, pp. 652–656.
6. Schonberger, V. M., and Cukier, K. Big data: A revolution that will transform how we live, work, and think.
7. The Penn Treebank Project. http://www.cis.upenn.edu/~treebank/.
8. The Stanford Natural Language Processing Group. http://nlp.stanford.edu/.

Chapter 5
Big Data and Cyber Foraging: Future Scope and Challenges

**Chhabi Rani Panigrahi, Mayank Tiwary, Bibudhendu Pati
and Himansu Das**

Abstract This chapter on "Big Data and Cyber Foraging: Future Scope and Challenges" mainly focuses on the development of cyber foraging systems. This chapter introduces the concept of cloudlets and the role of cloudlets in cyber foraging systems as well as discusses about the working and limitations of cloudlets. This chapter also explores the new architectures where the cloudlets can be helpful in providing Big data solutions in areas with less Internet connectivity and where the user device disruption is high. Different aspects in which the cloudlets can be used for managing and processing of Big data are also highlighted. The chapter then deals with different applications of cloudlets for Big data and focuses on the details of the existing work done with cyber foraging systems to manage different characteristics of Big data. Finally, this chapter explores the future scope and challenges of using cloudlets in the context of Big data.

5.1 Introduction

IBM states that in last two years 90 % of the data has been created from pervasive devices and social networks [29]. The increase of Big data including structured or unstructured data demands higher storage infrastructures. To minimize the cost of

C.R. Panigrahi (✉) · M. Tiwary
Department of Information Technology, C.V. Raman College
of Engineering, Bhubaneswar, India
e-mail: panigrahichhabi@gmail.com

M. Tiwary
e-mail: mayank09@gmail.com

B. Pati
Department of Computer Science and Engineering, C.V. Raman College
of Engineering, Bhubaneswar, India
e-mail: patibibudhendu@gmail.com

H. Das
School of Computer Engineering, KIIT University, Bhubaneswar, India
e-mail: das.himansu2007@gmail.com

© Springer International Publishing Switzerland 2016
B.S.P. Mishra et al. (eds.), *Techniques and Environments for Big Data Analysis*,
Studies in Big Data 17, DOI 10.1007/978-3-319-27520-8_5

storage, the data is distributed inside the walls of data centers. However, there is high requirement of architectures which can deal with the large volume of data. Distribution of data also facilitates parallel processing. The centralized storage is very slow and costly as compared to distributed environments. The development of cyber foraging [4] and invention of cloudlets solves the problem of deficiency of high processing resources among the users. As cloudlets are deployed geographically very near to its users, the cloudlets can be very helpful to check various characteristics of Big data such as high volume, value, velocity and veracity of Big data by appropriate methods. The cyber foraging systems if explored more in future can reshape the existing Big data architecture and help in more efficient management and processing the data in parallel in a distributed environment consisting of general mobile devices or intermittent devices. The cyber foraging systems express the power to take out the Big data from the walls of data centers to its users itself while at the same time ensuring high availability of the data and fault tolerant systems among intermittent devices.

5.1.1 Cyber Foraging

To create mobile devices with smaller size, lighter weight and longer battery life needs that their computational capabilities are required to get compromised. The growing market segments for mobile users mostly require computing and data manipulation facilities even beyond than those of the lightweight mobile computing devices consisting of longer battery life. Meeting such requirements is difficult for the service providers. One of the effective ways to deal with the same problems is to use cyber foraging systems. The main idea behind cyber foraging systems is dynamic augmentation of the computational resources for wireless mobile devices or computers is to exploit hardware infrastructure with wired connectivity. The computational resources are becoming very cheap and abundant. Therefore, the logic creates sense to "utilize" the computational resources being wasted in order to improve user experience for use of mobile-cloud services. The prices for desktop computers have fallen sharply in the recent years. The creators of cyber foraging systems by considering the for seeable future assumed that public spaces such as coffee shops and airport lounges being equipped with computational servers or with data staging servers for better customers support. The computational servers need to be connected to Internet through wired medium with high-bandwidth networks. Some of the commonly developed and used cyber foraging systems include [32]: Puppeteer [5], Spectra [11], Chroma [1], Scavenger [31], Goyal and Carter's system [6], Framework for Power Aware Remote Processing [12–15, 26], and Slingshot [27].

5.1.2 Cloudlets

The main aim of cloudlet is to bring the cloud closer to its users and is mostly viewed as data center in a box. The cloudlets can also be deployed over a micro data center [30]. According to the cloudlet architecture [28], cloudlets act as a shadow image of data centers and are also geographically very near to the users. The cloudlets can provide different kinds of services to the users and can also act as a central controlling device for wireless networks in three tire architecture represented by "mobile device—cloudlet—cloud" as represented in Fig. 5.1. Most of today's cloud applications use database to store structured data. The cloudlets can also have a mobile state where the cloudlet can itself run on general mobile devices. But, the most common state of a cloudlet is a powerful server deployed within a local proximity of the user in a LAN environment. The major benefits of cyber foraging systems include service access for its users in least or LAN latency environment. The cloudlets are connected to the Internet for synchronization with cloud and the degree of synchronization depends on the service being provided by the cloudlets. The core concepts of cloudlets show convergence of concepts from mobile computing and cloud computing.

The cloudlets act as backbone of present cyber foraging systems and are characterized by the following factors [28]:

- Only Soft state: The cloudlets mostly are cached state of cloud servers with no hard state. Soft state mainly signifies the flexibility in deployment of cloudlets over a specific category of devices in Internet of Things. The cloudlet is self-managing after its deployment or installation at the user site. The cloudlets can safely route the buffered data received from the mobile devices with high end security. This buffered data includes videos, photographs, data from sensors etc.

Fig. 5.1 General cloudlet three tier architecture

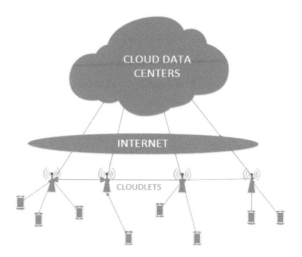

- Powerful, well connected and safe: The cloudlets are powerful and possess sufficient computational power (including sufficient CPU power, Physical memory etc.). These powerful cloudlets are mostly used for offloading computational intensive service requests to cloudlets instead of offloading it to the cloud for fulfillment of Service Level Agreement (SLA) for services being provided by the cyber foraged systems. The cloudlets are well connected to the Internet by different Internet gateways. The traffic demands from the cloudlet to cloud greatly depend on the type of service being provided by the cloudlet. The battery life of cloudlets depends upon the device on which the cloudlet is deployed. Integrity of cloudlet as a computing platform is assumed in a production-quality implementation this will have to be enforced through some combination of tamper-resistance, surveillance, and run-time attestation.
- Close at hand: The cloudlets are deployed geographically very near or locally to the proximity of the mobile users. The close at hand feature of cloudlets enables the cloudlets to provide low end-to-end latency and high bandwidth connection. The cloudlets are characterized by one-hop Wi-Fi device.
- Builds on standard cloud technology: The cloudlets represent the standard cloud architecture such as Amazon EC2 and Open Stack by encapsulating the code offloaded from mobile devices in to virtual machines (VMs). The functionality of cloudlet is dependent on the role of cloudlet.

The cloudlets can have two states and are described as follows:

- Fixed State: In this state, the cloudlet is deployed over non-portable high performance machines. The general resources of the high performance machines are characterized by high CPU power, high physical memory, and graphics processing units etc. This state has infinite power pool and requires very less power optimizations. This state characterizes the connectivity of cloudlets to the Internet through high bandwidth wired mediums and the Internet connectivity is very less prone to disruption. The cloudlets in this state are connected to the users through wireless access points in a LAN environment. The user end point devices are mostly Layer 2 devices and on the other end the cloudlets are connected to the Internet through routers and modems or mostly Layer 3 devices. The cloudlets in this state can also have dedicated storage systems such as Storage Area Network (SAN) or a Network Attached Storage (NAS). The selection of hardware for the cloudlets purely depends upon the type of service being provided and also the SLAs for different services. The cloudlets in this state can provide more services as compared to other states.
- Mobile or Moveable State: In this state the cloudlets are deployed over portable or movable devices which have the capabilities to provide services through Wi-Fi Hotspot to its users. The selection of devices greatly includes devices under Internet of Things with some constraints. The cloudlets can even be deployed over mobile devices or smart phones. In [20], authors introduced the cloudlet deployment over automotive environment. This automotive environment mostly includes moving automobiles which collect data from sensors connected across it. These devices mostly have finite power resources and optimization of power resources is required

Table 5.1 Evolution of hardware performance [10]

Year	Typical server		Typical handheld or wearable	
	Processor	Speed	Device	Speed
1997	Pentium® II	266 MHz	Palm Pilot	16 MHz
2002	Itanium®	1 GHz	Blackberry 5810	133 MHz
2007	Intel® Core™ 2	9.6 GHz (4 cores)	Apple iPhone	412 MHz
2011	Intel® Xeon X5	32 GHz (2 × 6 cores)	Samsung Galaxy S2	2.4 GHz (2 cores)
2013	Intel® Xeon® E5	64 GHz (2 × 12 cores)	Samsung Galaxy S4	6.4 GHz (4 cores)
			Google Glass OMAP 4430	2.4 GHz (2 cores)

in most of the cases. The mobile state of cloudlets creates opportunities for cloud service providers to extend their services in rural parts where there is shortage of high speed Internet Connectivity. The devices over which the cloudlets are deployed in this state are mostly intermittent in nature and proper architectures are required to reduce the disruptions for reliable services.

Table 5.1 represents a variety of devices which were commonly used in the form of wearable and mobile devices or smart phones. Table 5.1 represents that with passage of time, the mobile devices are getting more powerful processing power and battery life.

5.1.3 Big Data and Cloudlets

The cloudlets are placed very much closer to the proximity of the mobile or pervasive devices and provide services to the mobile users from one hop hotspot. The cloudlets can be used to check most of the Big data generated from the pervasive devices. The cloudlets can act as a layer between the pervasive devices and cloud for Big data. The cloudlets can act as central controller for the Big data or a proxy between the Big data generator and cloud. Acting as a middleman, the cloudlets can have much greater degree of affects on the Vs of Big data. The cloudlets have a wide range of advantages along with limitations also. Till date very less work has been done on the cloudlets for management and processing of Big data. Being very close to its users the cloudlets can even solve most of the real time problems in Big data. As the cloudlets act as cached state of actual data center, these are even able to provide services in absence of Internet (Fig. 5.2).

Fig. 5.2 Cloudlets in an
automotive environment [20]

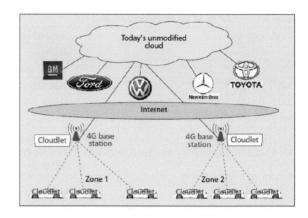

5.1.4 Contributions

In this chapter, different aspects in which the cloudlets can be used for managing
and processing of Big data are highlighted. The brief description of cloudlet and its
advantages and disadvantages in the context of Big data is also introduced. Specifically, this chapter focuses on the details about the existing work done with cyber
foraging systems to manage different characteristics of Big data. This chapter also
explores the new architectures where the cloudlets can be helpful in providing Big
data solutions in areas with less Internet connectivity and where the user device
disruption is high. In this chapter, the future scope and challenges for cloudlets in
context of Big data is also explored.

5.1.5 Organization

The remainder of this chapter is organized as follows. Section 5.2 introduces the
cloudlets in details which also includes working of cloudlets. Section 5.3 presents
the limitations of cloudlets. Applications of cloudlets with Big data is represented in
Sect. 5.4 where the existing work on cloudlets which affects the characteristics of Big
data is explained with the results and analysis. Section 5.5 discusses the effects of
cloudlets on data veracity. The future scope and challenges are explained in Sect. 5.6.
Lastly, Sect. 5.7 summarizes and concludes the chapter.

5.2 Working of Cloudlets

This section explains the cloudlets in details, alongwith how cloudlets work on existing cloudlets system and how cloudlets cope up to work in hostile environment.

5.2.1   Common Cloudlets Systems

In this section, different cloudlets systems are briefly described.

- **VMs on cloudlets**—In [23], authors introduced VMs to run over cloudlets. The cloudlets were given essential VMs to run based on the request from a variety of users. In [23], authors proposed Kimberlize as proof-of-concept to validate their proposed work. The Kimberlize created a VM overlay to run on the cloudlet. The VM overlay in Kimberlize is created by using the base VM, install script and resume script. The base VM consists of a base operating system installed and with minimal configuration. Kimberlize first launches the base VM and then executes the install script on the base VM. After execution of the install script, the VM is ready with applications for mobile device use. Lastly, the Kimberlize executes resume script after execution of install script to bring the state of application being run on the VM. Then, authors considered that VM as launch VM. After creating the launch VM, Kimberlize differences its memory and disk images with those of base VM for creation of VM overlay. Finally, the VM overlay is compressed and encrypted.

Figure 5.3 represents the runtime binding in Kimberlize. In Fig. 5.3, KCM represents Kimberley Control Manager. The KCM acts as a binding between the mobile

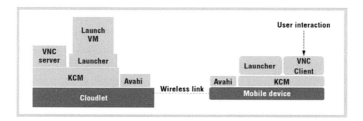

Fig. 5.3   Runtime binding in Kimberlize [23]

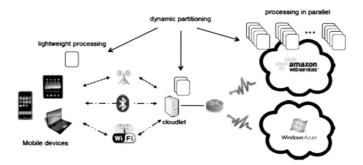

Fig. 5.4   General MOCHA architecture and dynamic task partitioning [24]

device and the cloudlet. Instances of KCM run on both mobile device and cloudlet. Service browsing and publishing is supported by KCM with the help of Avahi in Linux.

- **Cloud-Vision–Real-time face recognition using cloudlet**: In [24], authors proposed MOCHA as a proposed model for real time face recognition using cloudlet. The main challenge lies in task partitioning from mobile devices and offloading of the computational intensive operations. Figure 5.4 represents the MOCHA architecture. Here the mobile devices offload computational intensive operations to the cloudlets and the cloudlet dynamically partitions the computational intensive operations to be performed on mobile device itself, cloudlet, and the cloud also. In MOCHA, the cloudlets determine how to partition the computational intensive operations among multiple servers on cloud and itself to meet SLAs and optimize the Quality of Service (QoS) parameters. In MOCHA, the mobile devices acquire images and send to the cloudlets in raw form or pre-processed form. The cloudlet or mobile device does the pre-processing of the image and the cloudlet optimally offloads the processing to cloud or mobile device based on resource availability. After final processing of the image, the mobile device gets back its results.
- **Offload shaping for mobile devices**: In [9], authors tried to filter the data before being transmitted to the cloudlet for processing. Authors tried to propose solutions for blurry images, which posses to reduce the accuracy of computer vision algorithms. The cloudlets were programmed to run MOPED algorithm [9] for object detection services. The offload shaping filter was deployed on Google's glass, which used Sobel operator. The filter normally filters the blurry image frames for filtered transmission of images to the MOPED service on cloudlet.
- **Cognitive assistance using wearable devices**: In [8], authors tried to solve the problem of cognitive decline. Authors implemented a cognitive assistance system Gabriel which runs on cloudlets and gets computational intensive operations and tasks from wearable devices such as Google's glass. Figure 5.5 represents Gabriel architecture which runs on a cloudlet deployed in very close proximity of the users. In Fig. 5.5, the Gabriel architecture works by implementing different engines of cognitive assistance on different VMs. Coarse-grained parallelism is trivially exploited because of no shared state across multiple VMs running different cognitive assistance engines.

5.2.2 Role of Cloudlets in Hostile Environment and Areas with Less Internet Connectivity

In [22], authors extensively studied the performance of cloudlets in hostile environment. In hostile environment, there is limited connectivity or no connectivity from the user to the cloud. Examples of hostile environment include areas where recovery is in action due to any terrorist activities or disasters or hazards, theater of military operations. Another example of hostile environment includes public region with

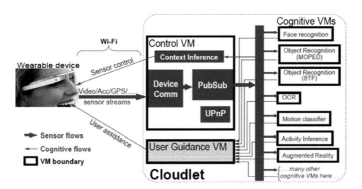

Fig. 5.5 Gabriel architecture for cognitive assistance [8]

well connected Internet space in early stage but Internet connection disruptions due to cyber attacks.

The cloudlets were originally developed to solve the problems having constraints from high latency. The cloudlets tend to reduce the end-to-end latency drastically in most of the considered cases. But, the availability or deployment of cloudlet in the physical proximity of the users, bring higher degree of advantages to its users other than reduced end-to-end latency.

The cloudlets act as cached state of the data center and are able to provide services to its users even at the hostile environment. The power of cyber foraging systems which helps the cloudlets to provide services during hostile environment can have a very higher degree of affect on the economy of developing nations. Most of future scope of cloudlets is discussed in later sections. The cloudlets can even be deployed by the data center brokers [18]. In [18], authors introduced CAB as a model in which the cloudlets act as agents of cloud brokers and the main aim of cloudlet is to provide real time services to the users to specifically meet the SLA for the services being provided by the cloudlets. In [18], authors also introduced a pricing scheme for CAB and price optimization algorithms for the cloudlets which aims to benefit the data center brokers. The price optimization is solely based on the resource scheduling by the cloudlets. In [18], authors did not consider mobile cloudlets, where as they considered fixed state cloudlets with high or powerful computing resources.

Figure 5.6 represents a hostile environment situation in which the cloudlets form a mesh topology network, start communicating with themselves and provide services even if the Internet connectivity is weak or very weak with the actual cloud. The performance of the cloudlets network entirely depends on the applications state optimizations or in other words, the performance of cloudlets is dependent on the type of service being provided by the cloudlet.

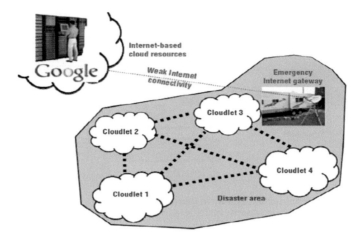

Fig. 5.6 Cloudlets meshed connectivity during hostile environment [22]

5.2.3 Mobile Cloudlets and Mobile Applications

As the state of cloudlets can also support mobile devices such as smart phones, it is very important to consider "can mobile cloudlets support mobile applications". In [16], authors examined the fundamental mobile cloudlet properties that suggest whether and when a mobile cloudlet can provide services for mobile applications. The results which authors analyzed suggest that:

(1) The more frequent mobile devices are discovered, the more amount of computational resources an initiator or user can access.
(2) Disruption of connection between the mobile device and cloudlets has a little effect over the performance in long run.
(3) Authors introduced upper bound and lower bound on computing capacity and computing speed of a mobile cloudlet, which helps the users to decide whether to offload computational intensive tasks to the mobile cloudlet.

In this work, authors proposed mobile cloudlets as set of mobile nodes, which can provide mobile services and at the same time, can also be service request initiator. Whenever the cloudlet receives service requests from an initiator node, the task gets divided among all the nodes in the cluster and ultimately the initiator can conserve energy and speedup computations. Both traces and results analysis together determine that mobile cloudlet size follows negative exponential growth with time. Where the mobile cloudlet size or the number of mobile nodes in cluster, which form the cloudlets. The expected lifetime grows linearly with increase rate 1 and the expected reachable time of a cloudlet node grows linearly with slope $E(T_C)/[E(T_C) + E(T_I)]$, where $E(T_C)$ and $E(T_I)$ are the expectations of contact and inter contact time. $E(T_C)/[E(T_C) + E(T_I)]$ indicates the probability of reach ability of initiator to the mobile cloudlet for task dissemination and retrieval and thus can be used as an

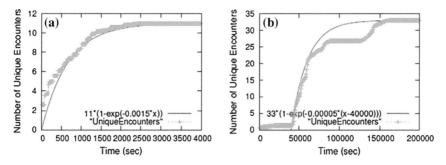

Fig. 5.7 Mobile cloudlet sizes follow negative growth exponentially with time in two different experiments [16]. **a** Exp1. **b** Exp2

important parameter in deciding whether to offload task to mobile cloudlet so that SLA meets (Fig. 5.7).

5.3   Limitations of Cloudlets

This section represents the limitations of the cloudlets. The limitations are analyzed in terms of probability of cloudlet access and probability of successful task execution, average speed of task execution and average tasks executed.

5.3.1   *Cloudlet Network Model*

In [16], authors assumed movement of mobile device in a network μ_m consisting of m cloudlets. The network gets partitioned into Voronoi diagram with m number of voronoi cells. Each voronoi cell is characterized by the presence of one cloudlet node. Let us consider a computational task on a mobile device that requires C instructions. Let S_i be the computational speed which is represented in instructions per time unit (usually a second) of the cloudlet C_i where, i = 1, 2, …, m. Now the task consisting of C instructions takes C/S_i number of unit time slots to perform the total computation for the service request received from the mobile user. This model denotes B as the network bandwidth. If a mobile user requires to send D amount of task data in bytes to the cloudlet, whereas the cloudlet needs to send K bytes of data to the mobile user. Normally the time taken to transmit the data is represented as D/B and K/B to send and receive data respectively. This model defines task completion time as i = C/S_i + (D+K)/B. The task completion time is defined as the sum of the task computing times at a given cloudlet i and task transmission time.

5.3.2 Cloudlet Connection Model

In [21], authors proposed a model which considers the access of a mobile device to a given cloudlet when C_i is within the range of the mobile cloudlet $1 \le i \le m$. Otherwise, the mobile device seems to be disconnected from the mobile cloudlet. The locations of the mobile device and mobile cloudlet is given by X(t) and $X_{Ci}(t)$. Let r be the range up to which the mobile cloudlet accepts connections and is capable of providing the service with no interruptions. At any time t, the connection is available if and only if $\|X(t) - X_{Ci}(t)\| \le r$, here $\|.\|$ is the Euclidian norm in two dimensions.

Definition 1 The mobile device and the mobile cloudlet C_i connection time T_C is defined as follows:

$$T_c^i \equiv inf_{t>0}\{t : \|X(t) - X_i(t)\| > r\} \tag{5.1}$$

Definition 2 A given process $\{\phi(t), 0 < t < \infty\}$, (stochastic process) and given state space $\{0, 1\}$. If a mobile device can connect to a cloudlet at time t, $\phi(t) = 1$, otherwise $\phi(t) = 0$. The process $\phi(t)$ assumes that the state 0 and 1 alternately as shown in Fig. 5.8. The process $\phi(t)$ is defined as alternating renewal process.

Theorem 1 *The cloudlet access probability is given as follows:*

$$CA = \sum_{i=1}^{m} \frac{\mu_{T_C}^i}{\{\mu_{T_C}^i + \mu_{T_I}^i\}} \tag{5.2}$$

where, μ_{TC}^i is defined as the expectation of connection time T_C^i for mobile device and μ_{TI}^i is defined as the expectation of inter-connection T_I^i for mobile cloudlet C_i.

Theorem 2 *For a cloudlet, the task success rate is denoted by SR and is defined as follows:*

$$SR = \sum_{i=1}^{m} CA^i \left(1 - \int_0^{\delta_i} [1 - F_{T_c}^i(x)] / \mu_{T_c}^i dx \right) \tag{5.3}$$

Here, CA^i is the defined as the cloudlets access probability and is given as follows:

Fig. 5.8 Alternating renewal process graph for connection and inter-connection process of a mobile device and a mobile cloudlet C_i [16]

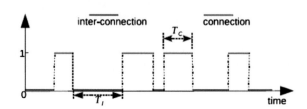

$$CA^i = \mu^i_{T_C} \div \left\{ \mu^i_{T_C} + \mu^i_{T_I} \right\} \tag{5.4}$$

Theorem 3 *The tasks executed in average over unit time t is denoted by N(t), which satisfies the following:*

$$E(N(t)) = \sum_{i=1}^{m} \lfloor \frac{E(N^i_C(t))\mu^i_{T_C}}{\delta_i} \rfloor \tag{5.5}$$

Here, $E(N^i_C(t))$ *is defined as the connections existing between a mobile device user and a mobile cloudlet C_i limited by time t.*

Theorem 4 *The speed of average task execution is denoted by CS and is defined as follows:*

$$CS = \lim_{t \to \infty} \frac{E(N(t))}{t} = \sum_{i=1}^{m} \frac{CA^i}{\delta_i} \tag{5.6}$$

All the above theorems try to put limitations on the cloudlets considering the cloudlet access probability, probability of successful task execution, speed of average task execution and average tasks executed. These factors become very important to consider when the cloudlet will be used to put affects on the different Vs of Big data. These factors are also important when the cloudlets need to run map-reduce programs and to optimize the battery life of the mobile devices. The cloudlets with fixed state are deployed over powerful servers, in the physical proximity of the users. The above defined theorems and definitions vary for fixed state cloudlets.

5.4   Applications of Cloudlets for Big Data

This section explains some of the existing applications of cloudlets over Big data. The cloudlets can be very helpful in checking the different aspects of Big data. As Big data is the data collected effortlessly generated from different sources such as the data generated from sensors, CCTV cameras, wearable devices, and pervasive devices etc. As the cloudlets are deployed in the physical proximity of the Big data generators, it can directly process the data on site and then send the processed data or the results to the cloud. The presence of cloudlets in the physical proximity of the Big data generators adds an extra layer of filter having computational power.

5.4.1   GigaSight

From Internet of Things, vast amount of data is generated from real time machines [3]. These machines mostly include sensors and surveillance cameras. In the year

2013 in USA, it was analyzed that for every 11 persons, there exists one or more than one surveillance cameras [2]. The emergence of wearable devices under Internet of Things such as GoPro, Google Glass, etc. is trying to shape a future in which the video cameras will become body-worn and most commonly used and essential part of human life. These cameras are one of major player in generating the Big data. The data generated from these cameras are unstructured in nature and require complex processing systems for it. Maximum cost an organization faces is for the storage of the generated video streams from the video cameras. NSF workshop on Future Directions in Wireless Networking, 2013 analyses that "It will soon be possible to find a camera on every human body, in every room, on every street, and in every vehicle" [19]. Today's storage architecture which store the video streams from the cameras are deployed very near to the video cameras. The videos generated from these devices are very much rich in content and value as compared to the data generated from sensors.

GigaSight is an effective architecture which is based on reverse form of Content Delivery Network (CDN) and primarily focuses on Internet based searchable repository for variety of video streams. The category of video streams over which the GigaSight primarily focuses is crowd-sources video streams. Some of the hypothetical use cases for video streams from crowd sources can be (1) Marketing: Which wing attracts the people most? Which Animal is showing unusual behavior? etc. (2) Surveillance based on computer Vision: Locating people, pet, objects, etc. from different video streams, searching a thief from a list of thief from the surveillance video streams, etc.

In the present architecture, when these challenges are tried to address the video stream needs to be transmitted to the cloud where the video streams will be processed with high end powerful computing devices. But, this architecture possesses a variety of network challenges.

The improper utilizations of ingress resources that act as communication channel between the video source generator and the cloud mostly results into traffic congestions. In GigaSight system, the video source generators transmit the video stream in real time to the cloudlet in the physical proximity. The cloudlet buffers the video stream and runs computer vision algorithms on the video streams buffered then the results collected after running the computer vision analytics is sent to the cloud, which reduces the traffic to a great extent as compared to the size of the video source. The results which are sent from the cloudlet to the cloud mostly include recognized objects, recognized faces, time of detection, space of detection, etc. These results are also followed by the metadata such as object name, person name, time of detection, etc.

The selection of computer vision analytics algorithms that run on the cloudlets purely depends upon the problem being analyzed. The video streams which are sent to the cloudlet can also receive blurry frames which can reduce the accuracy of the computer vision algorithms, so offloading also needs to be shaped as described earlier in Sect. 5.2.1. GigaSight also preserves the privacy for better way of management. To better preserve the privacy, the video streams need to be modified or altered based on the requirements of the organization. This modification or alteration of videos needs

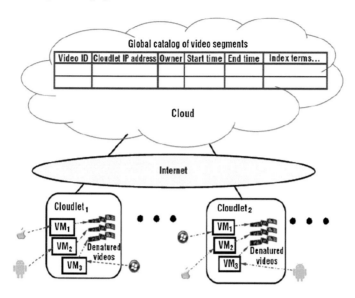

Fig. 5.9 GigaSight architecture deployed over cloudlets [3]

to be automated. Authors of GigaSight term the automation process for preserving privacy as denaturing (Fig. 5.9).

Denaturing can also affect the value of the video stream being processed. Denaturing is highly sensitive to the context in which the GigaSight is being used. Denaturing carefully analyzes the frames and tries to process the frames based on requirements and mostly require high computational resources. Denaturing may involve blurring or completely removing the privacy constraints appearing in the video streams. Some of the common use cases of denaturing video streams include removing specific faces, objects, or personal scenes. The denaturing of data not only involves modifying the contents of a video stream but may also focus on metadata modification. This is done by taking the objects that needs to be removed from the scene, extracting features from the objects and training the machine classifiers.

Whenever the denaturing processing systems recognize an object, similar process occurs for the recognized objects i.e. features are extracted from the newly recognized objects and testing the trained machine classifier. This is one of the methods for denaturing a video stream, but there can be many methods which can do the same. Denaturing of data is not only associated with cloudlets, but also with those systems where cloudlets will be used for data analysis at the physical proximity of the data source generators. The above defined method for denaturing a video stream requires optimization methods for feature selection, training of the machine classifier and testing. The optimizations methods are mostly context sensitive. The denaturing of video streams have higher degree of affects over the value of Big data.

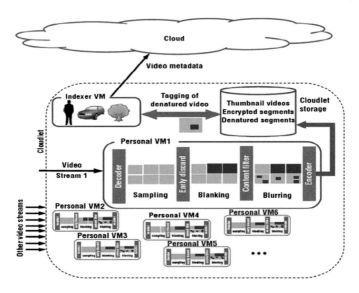

Fig. 5.10 Gigasight deployed over cloudlets [3]

In Fig. 5.10, the GigaSight model is deployed over the cloudlet. The figure also shows multistep pipeline deployment of denaturing engine. Figure 5.10 also shows the implementation of video stream denaturing engine on a VM.

The video stream denaturing engine is implemented over on a VM (Personal VM) over a cloudlet. The denaturing algorithms run on the specified VM based on clients preferences, whose data and processes are separated from other VMs. The user mobile device or the video stream generator and the personal VM has only access to the original video stream. There are multiple personal VMs over the cloudlet, whcih have denaturing engine implemented over them.

In the initial phase, a part of original video i.e. set of frames are used by the denaturing. The denaturing engine produces two outputs, one is "thumbnail video" and encrypted version of the original video. The thumbnail video is used to represent the overview of video content and is mainly used for search operations and indexing. Both of the outputs are stored on the cloudlet storage. While a search operation is performed, over the thumbnail video and results a section of intrest for the full-fidelity video then its respective personal VM is requested to decrypt and denature the particular segment required. After the decoding phase, the filters which are used for the privacy of metadata are applied. This is a binary process in which the frame is either completely allowed or is completely blanked or not allowed. After this the content based filters are applied which is based on the user preferences. Figure 5.11 shows the frames of a denatured video.

The indexing as shown in Fig. 5.10 is performed on a separate VM and the indexing process runs as a background activity. For searching requests, custom code must be written which can directly examine the denatured video and give appropriate results.

Fig. 5.11 Example of
denatured video frames [3]

The performance or throughput of the indexing VM depends on the total number of objects that are to be detected. In other words, the indexer's performance depends on the context and the type of service being provided by the cloudlet. In real world situations, the number of objects can be very high, so authors in [3], proposed an approach to use machine classifiers for those object which occur frequently in the context.

GigaSight is based on a two step hierarchical workflow for its searching operations. In the first step a user performs a SQL query search, which may consists of searching about the metadata. After the search or query is executed, the user gets the reply in the form of list of video segments and respective thumbnailed video. The second step is based on filtering out the search to get the actual content in a more relevant set. This process is computationally intensive and Gigasight ensures to run them in parallel to increase performance.

GigaSight also models the automotive cloudlet architecture in which the cloudets are deployed over moving vehicles or base towers. Under this model, the video cameras are deployed over the moving vehicles which can be used to ensure higher degree of road safety.

For example if the computer vision algorithms are able to track dead animals, fallen trees, broken roads during disasters and accidents, then the cloudlets deployed over the moving automobiles can easily track their location and warn other vehicles for safety. This communication of warning message takes place through the cloudlets at the base towers, which receive the warning messages from one of the vehicle and brodcast this message to all other vehicles whose cloudlets can be reachable from the base towers cloudlets. These video streams an further be used by survey agencies to do different kinds of automated surveys for the traffic management systems. Todays automobiles are griped with high data rate sensors over different parts of the vehicles.

These sensors usually transmit the data to cloud, where the data is analysed and results containing preventive measures are sent back to the automobile via Internet. But the chances of failure of this model is high when the vehicles are in such areas where the Internet connectivity is very disruptive in nature. The other challenges that

Fig. 5.12 Users per cloudlet versus frames processed per second per user [3]

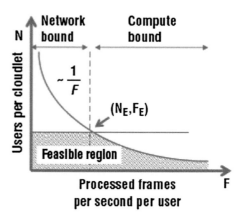

the cloud service providers face is regarding the storage of the high velocity data from the sensors of the automobiles. The cloudlets when deployed over the vehicles can do the processing of data on the site itself and provide results to the vehicles in real time. The cloudlets can even work in those environments where their is disruptive Internet connection.

The graph as shown in Fig. 5.12 represents the tradeoff between the frame rates being processed per user and number of users per cloudlet. In Fig. 5.12, the shaded region facts that the range for feasible elements based on processing power and fixed network. Here N is the number of users and F is the number of frames processed per user.

5.5 Big Data Veracity and Cloudlets

This section explains the different problems which can be solved at the user end or the cloudlets without transmitting the computational data to the cloud. The unstructured data constitute a major part of the Big data generated and it posses threats internally to the administrators to secure data. This section explains the problems which are in the context of data veracity.

5.5.1 Malware Detection in Unstructured Data Using Cloudlets

Malwares are found in the unstructured data and securing the big unstructured data from malwares is a big challenge for cloud storage providers. The increasing number of malware signatures makes it very difficult to scan the whole database for malwares. In [25], authors proposed a malware detector based on map-reduce programming approach. The map-reduce malware detector engine uses Graphics Processing Unit (GPU) at the back end for parallel processing of the data. The map-reduce programs run in parallel and along with the parallel map-reduce programs, the GPU parallelism provides second level of parallelism for the map-reduced based malware detection engine. Examples of unstructured data include videos files, mp3 files, compressed files (such as gzip, bzip, zip, rar, iso, etc.), executable files etc.

There are two methods for malware detection; (1) Signature based (2) Behavior based. In the signature based malware detection, the files contents are scanned against a set of malware signatures whereas in the behavior based malware detection, the behavior of files are monitored against a set of pre defined behavior. The signature based malware detection is highly computational intensive and hence solved by map-reduce approach. The virus signatures from the Clam AV [25] are divided into three main categories:

- **Basic Signatures**: Basic signatures are in the form of hexadecimal strings. These signatures are scanned with the whole content of the file.
- **MD5 Signatures**: These signatures are in the form of MD5 checksums. These signatures are matched with the MD5 checksums of a file instead of matching the whole content of the file.
- **Regular Expression Signatures**: These signatures consists several forms of wild cards which are evaluated with the inspected file. The CPU overhead is very high in case of basic signatures and regular expressions.

Map-Reduce AV algorithm (GPU based)

In the mapping phase, the input is the file to be scanned and the size of the input is based on blocks of size 64 MB. The mapper breaks the input block into sentences and grouped into array of two dimension form. Then the 2D array is sent to the GPU using asynchronous streaming process among a single block of size 64 MB. The malware signatures are cached in the distributed cache, which is also grouped similarly into 2D array and sent to the GPU's read only cache. During the mapping phase, the mapper receives the block as value and the name of the file as key. The GPU kernel is called in the mapper which actually performs the pattern matching.

In [17], authors implemented two pattern matching algorithms—Wu-Manber's algorithm and Bloom Filter algorithm over GPU and evaluated their performances. After the GPU finishes its execution, the result is sent in binary form to the CPU, where 1 indicates matching of the signatures with the contents of the file and 0 indicates no match. Then the mapper emits the file name as key and the binary result from GPU as value to reducer. The reducer checks every key for values 1, and emits the same key and value as "contains malware contents" to file system to be written as results. One of the major limitations of this work is that no asynchronous activity can be established among different mappers to reduce the bottleneck of bandwidth of PCI slot where the GPU card is attached physically. In this work, authors tried to evaluate the performance of CPU by using java's multithreading, which can help in complete evaluation of CPU's cores and a fair comparison can be made between the hadoop cluster without GPU cards and with GPU cards.

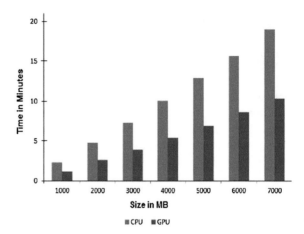

Fig. 5.13 Time comparison graph for Bloom Filter algorithm

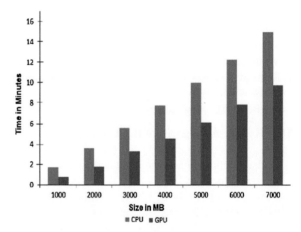

Fig. 5.14 Time comparison graph for Wu-Manber algorithm

Figures 5.13 and 5.14 represent the time comparison graphs for bloom filter algorithm and Wu-Manbers algorithm respectively. In this analysis, authors evaluated the proposed mapreduce malware detector engine by giving different inputs consisting of unstructured data and 15000 malware signatures from Clam AV.

Figure 5.15 represents an overall comparison between the Bloom Filter algorithm and Wu-Manbers algorithm, where the Wu-Manbers algorithm for pattern matching seems to be a better solution when the size of malware signatures is high.

In cyber foraging, whenever a user uploads a file to cloud via cloudlet, the cloudlet will evaluate the file for malware and secure the data before it is uploaded to the cloud. For this problem, cloudlets with high computational resources are required with a GPU card. After the cloud performs the evaluation of the data, then the cloudlet uploads the file to the cloud storage. In this context, the cloudlet acts as a Firewall with main aim to secure the data.

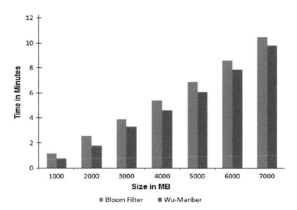

Fig. 5.15 Time comparison graphs for Bloom Filter algorithm versus Wu-Manber algorithm

5.5.2 Steganalysis at Cloudlets

A major percentage of today's unstructured data constitute videos and images. Steganography is the art of hiding information, which embeds secret messages into images, mp3 files, etc. using a variety of methods. The steganography also possess threat to data integrity which needs to be addressed at the earliest. In today's Big data context, the high volume of data creates challenges for steganalysis. If the steganalysis is performed at the earliest at the cloudlet, it would benefit a lot to the data center administrators.

There are a variety of methods for steganalysis. In [6], authors used a three phase method for the same. In [6], authors proposed steganalysis engine using GPU's to accelerate the speedup of execution and to cope up with the high volume of data. In [6], authors extracted features from the transform domain and trained the machine classifiers for detecting an image as stego or non-stego based on the appropriate features. The steganalysis engine is based on three phases as shown in Fig. 5.16.

The first phase of steganalysis engine is the image estimation. In this phase, order statistics filter are applied to the original image and thereby producing five estimated images. These estimated images are considered to have less noise as compared to the original stego image. The five order statistics filters applied are median filter, min & max filter, mid filter and alpha trimmed mean filter. In the second phase, features are extracted from the original image and the five estimated image. The authors generated a set of 255 features from the input image using first order statistics

Fig. 5.16 Steganalysis engine

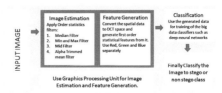

features from [7]. The first order statistics includes three functions, these functions have several attributes, when these attributes are applied over the five estimated images and the original image then a set of 255 features are generated. In the third phase, the features generated from the second phase are used to train the neural network machine classifier. The features generated for input image are normalized and given as input to the neural network. The neural network is now trained and ready to operate for final classification of the input image to stego or non stego.

For training the neural network, features from stego and non stego image are used. The feature extraction process is the most computational intensive operation and is thus offloaded to the GPU for execution by implementing parallel GPU kernels. Final classification results show higher accuracy of classification against the extracted features. Figure 5.17 shows the convergence and performance graphs for the artificial neural network. This graph represents the performance of the neural network during training phase of the classifier. Figure 5.18 shows the time comparison graphs for the performance of steganalysis engine without GPU card and with GPU card attached. The steganalysis engine with GPU cards attached shows higher degree of performance. The results analysis depicts that there is not much difference between the overall speedup of the GPU against CPU this is because of two main factors: one the use of JCUDA which adds extra overhead of the Java Native Interface (JNI) and the use of multiple GPU kernels during the feature extraction phase, which restricts the asynchronous activity of the GPU cards to some extent.

When this steganalytic engine is deployed over the cloudlets, the cloudlets ensure only non stego images to transmit to cloud. In this context also, the cloudlet acts as a firewall for Big data and ensures high end security for the data that is transmitted to the cloud for storage.

Fig. 5.17 Training performance and convergence graph

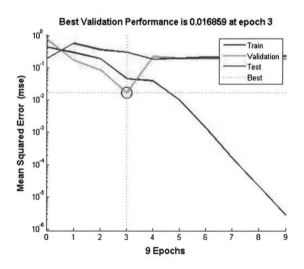

Fig. 5.18 Time comparison graphs for CPU and GPU

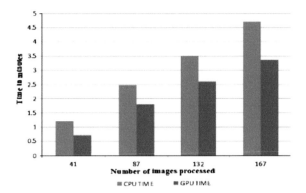

5.6 Future Scope and Challenges

This section explains the future vision that the cloudlets can enable for future generation Big data computing and the challenges faced by the cloudlets.

5.6.1 Future Scope for Cloudlets

The cloudlets are the major elements of cyber foraging and enable on site computations. The cloudlets are deployed on the physical proximity of the users and are able to cache the services from the cloud. The cloudlets and cyber foraging models and architectures are trying to reshape the existing cloud models. There are a variety of fields in which the cloudlets can be used to cope up with the challenges of the Big data. Some of those are explained and highlighted as follows:

1 Rural Uses—The cloudlets are capable of providing services during hostile environment. In rural areas of developing countries, the reach of Internet is not so developed and the existing research to advance the growth of rural areas depends mostly on high speed Internet. In rural areas, the cloudlets can be deployed to the sensors in agriculture fields, which can provide services there. The cloudlets do not require high computational facility at the fields, where as they can be adjusted by deploying them to mobile devices. The Hospitals in rural India can use Body Area Networks sensors for constantly monitoring the health of rural people, the cloudlets if deployed in the physical proximity of the sensors, can have benefits over real time problems.

2 Urban and Semi Urban Uses—The cloudlets can be used greatly to develop the health technology for urban and semi urban health care systems. The cloudlets can help in real time health monitoring systems, on site diagnosis for a variety of diseases using trained machine classifiers. The automotive model of cloudlets can enable them to take the computation to the edge of society. The automotive cloudlets can further be useful for auto driving vehicles, ensuring higher road safety, automated emergency calls to the nearest health care systems, etc.

3 Learning Systems—The mobile cloudlets can become helpful to create local data repository managed by cloudlets for fault tolerance and high availability of data. This property of cloudlets can enable them to cope up with the high volume problems of Big data by enabling "your data with you" model.

5.6.2   Challenges with Cloudlets

The cloudlets are deployed in the physical proximity of the users or Big data generators. The deployment of cloudlets near to the users also adds extra cost to the cloud service providers. The cloudlet also needs Internet connection and the use of Internet traffic by cloudlet depends upon the type of service being provided. The latest research on software defined approaches such as Software Defines Networks need an integration of current research with cyber foraging systems. The software defined approaches for cloudlets can help the cyber foraging systems to highly optimize the resources of cloudlets and provide reliable services to the users and at the same time help in increasing the profits at the cloudlet service providers.

The mobile cloudlets have limited battery health or energy and hence there is the need of highly optimized techniques for use of cloudlets services which can reduce the energy consumption for the cloudlets. There can be a lot of research on the states of cloudlets. There is a specific category of devices under Internet of Things, which can be used for cloudlets deployment. The cloudlets are self managed and require very less management. Currently, very less work has been done to reduce the management issues for cloudlets, which need to explore in the future for adding managerial and deployment benefits to the cloudlets. The costs of cloudlets also depend upon the type of facilities and services being provided by the cloudlets. The cloudlets can be used to control the high velocity of Big data such as congestion controll and appropriate bandwidth allocations.

5.7   Conclusion

The cloudlets are one of the prominent and the most important element of the today's cyber foraging systems. The cloudlets are able to provide services during hostile environment. As the cloudlets are deployed over the physical proximity of the users or Big data generators, the cloudlets can be used as Big data firewalls, which can filter the content of Big data generated before uploading the data to cloud. This nature of cloudlet has a great effect over the veracity factor of the Big data. The mobile cloudlets along with the fixed state cloudlets can be used to create local data repositories to cope up with the high volume of Big data. The cloudlets can also have effect on the value of Big data as seen in the GigaSight project. This chapter also reveals the future scope and challenges for cloudlets and cyber foraging systems as well as limitations of cloudlets.

References

1. Balan, R. (2003). Tactics-based remote execution for mobile computing. In *Proceedings of the 1st international conference on Mobile systems, applications and services (ACM)* (pp. 273–286).
2. Banerjee, S., & Wu, D. O. (2013). Final report from the NSF Workshop on Future Directions in Wireless Networking. National Science Foundation, (pp. 1–37).
3. Barrett, D. (2013). One surveillance camera for every 11 people in Britain, says CCTV survey. *The Telegraph*. Retrieved July 10, 2013, from www.telegraph.co.uk/technology/10172298/One-surveillance-camera-for-every-11-people-in-Britain-says-CCTV-survey.html.
4. Bohez, S., Verbelen, T., Simoens, P., & Dhoedt, B. (2014). Allocation algorithms for autonomous management of collaborative cloudlets. In *Proceeding of 2nd IEEE International Conference on Mobile Cloud Computing, Services, and Engineering (MobileCloud)* (pp. 8–11).
5. de Lara, E., Wallach, D. S., & Zwaenepoel, W. (2001). Puppeteer: Component-based adaptation for mobile computing. In *Proceedings of the 3rd Conference on USENIX Symposium on Internet Technologies and Systems (USENIX Association)*.
6. Goyal, S., & Carter, J. (2004). A lightweight secure cyber foraging infrastructure for resource-constrained devices. In *6th IEEE Workshop on Mobile Computing Systems and Applications (IEEE)* (pp. 184–195).
7. Grossman, R., Gu, Y., Sabala, M., & Zhang, W. (2008). Compute and storage clouds using wide area high performance networks. *Future Generation Computer Systems* (pp. 179–183).
8. Ha, K., et al. (2014). Towards wearable cognitive assistance. In *Proceedinghs 12th Initial Conference on Mobile Systems, Applications, and Services, Bretton Woods, NH* doi:10.1145/2594368.2594383.
9. Hu, W., Amos, B., Chen, Z., Ha, K., Richter, W., Pillai, P., Gilbert, B., Harkes, J., & Satyanarayanan, M. (2015). The case for offload shaping. In *Proceedings of the 16th ACM International Workshop on Mobile Computing Systems and Applications (HotMobile'15), New York, NY, USA* (pp. 51–56).
10. Flinn, J. (2012). *Cyber foraging: Bridging mobile and cloud computing via opportunistic offload*. California: Morgan & Claypool Publishers.
11. Jason Flinn, S. P., & Satyanarayanan, M. (2002). Balancing performance, energy, and quality in pervasive computing. In *22nd IEEE International Conference on Distributed Computing Systems*, (pp. 1–10).
12. Kafer, G., Haid, J., Voit, K., & Weiss, R. (2002). Architectural software power estimation support for power aware remote processing. In *Proceedings of the Fourteenth IASTED International Conference on Parallel and Distributed Computing and Systems (PDCS 2002)*, doi:10.2316/Journal.202.2004.2.204-0553.
13. Kafer, G., Haid, J., Hofer, B., Schall, G., & Weiss, R. (2001). Framework for power aware remote processing: Design and implementation of a dynamic power estimation unit. In *Proceedings of the IEEE Symposium on Wearable Computers (ISWC 2001)*.
14. Kafer, G., Haid, J., Schall, G., & Weiss, R. (2001). The standard power estimation interface for software components. In *Proceedings of the Workshop on Mobile Computing (TCMC 2001)*.
15. Kafer, G., Haid, J., Schall, G., & Weiss, R. (2001). Framework for power aware remote processing: The dynamic power estimation process. In *Proceedings of the Workshop on Mobile Computing (TCMC 2001)*.
16. Li, Y., & Wang, W. (2014). Can mobile cloudlets support mobile applications? In *Proceedings of IEEE INFOCOM* (pp. 1060–1068).

17. Mustafi, J., Mukherjee, S., & Chaudhuri, A. (2002). Grid computing: The future of distributed computing for high performance scientific and business applications. In *Lecture Notes in Computer Science, International Workshop on Distributed Computing* (Vol. 2571, pp. 339–342). Berlin: Springer.
18. Panigrahi, C., Tiwary, M., Pati, B., & Misra, R. (2015). CAB: Cloudlets as agents of cloud brokers. In *International Conference on Advances in Computing, Communications and Informatics (ICACCI)*, ISBN 978-1-4799-8790-0, (pp. 381–386).
19. Panigrahi, C. R., Tiwari, M., Pati, B., & Prasath, R. (2014) Malware detection in big data using fast pattern matching: A hadoop based comparison on GPU. *Mining Intelligence and Knowledge Exploration* (pp. 407–416).
20. Satyanarayanan, M., Schuster, R., Ebling, M., Fettweis, G., Flinck, H., Joshi, K., et al. (2015). An open ecosystem for mobile-cloud convergence. *IEEE Communications Magazine, 53*(3), (pp. 63–70).
21. Satyanarayanan, M., Simoens, P., Xiao, Yu., Pillai, P., Chen, Z., Ha, K., et al. (2015). Edge analytics in the internet of things. *IEEE Pervasive Computing, 14*(2), 24–31.
22. Satyanarayanan, M., et al. (2013). The role of cloudlets in hostile environments. *IEEE Pervasive Computing, 12*(4), (pp. 40–49).
23. Satyanarayanan, M., et al. (2009). IEEE Pervasive Computing. *The case for VM-Based cloudlets in mobile computing, 8*(4), (pp. 14–23).
24. Soyata, T., Muraleedharan, R., Funai, C., Kwon, M., & Heinzelman, W. (2012). Cloud-Vision: Real-time face recognition using a mobile-cloudlet-cloud acceleration architecture. In *IEEE Symposium on Computers and Communications (ISCC)* (pp. 59–66).
25. Tiwary, M., Priyadarshini, R., & Misra, R. (2014). A faster and intelligent steganography detection using graphics processing unit in cloud. In *International Conference on High Performance Computing and Applications (ICHPCA)* (Vol. 1, No. 6, pp. 22–24).
26. Voit, K. (2003). Implementation of a Transparent Code Migration Unit for Power Aware Remote Processing using JAVA.
27. Ya-Yunn, S., & Flinn, J. (2005). Slingshot: Deploying stateful services in wireless hotspots. In *Proceedings of the 3rd international conference on Mobile systems, applications, and services(ACM)*.
28. Cloudlets Based Mobile Computing. Retrieved June 10, 2015, from http://elijah.cs.cmu.edu/ (Online).
29. IBM. Big data at the speed of business, February 2015. Retrieved Feb 19, 2015, from http://www--01.ibm.com/software/data/bigdata/ (Online).
30. Micro Data Center Design. Retrieved June 10, 2015, from http://www.inveneo.org/designchallenge/.
31. Scavenger. Cyber foraging, June 2010. Retrieved June 10, 2015, https://code.google.com/p/scavenger--cf/ (Online).
32. WikiPedia. Cyber foraging, June 2015. Retrieved June 10, 2015, from http://en.wikipedia.org/wiki/Cyber_foraging (Online).

Chapter 6
Parallel GA in Big Data Analysis

**Santwana Sagnika, Bhabani Shankar Prasad Mishra
and Satchidananda Dehuri**

Abstract As the scope of computation is extending to domains where large and complex datasets are needed to be dealt with, it has become a very useful approach to sub-divide the tasks and perform them in parallel, which leads to a significant reduction in the processing time. On the other hand, evolutionary and swarm-based algorithms are rapidly gaining popularity to solve complex problems. However, these methods consume a lot of time in solving problems. Hence, parallelization of evolutionary algorithms proves to be beneficial in solving intensive tasks within a feasible execution time. This chapter describes the parallelization issues in Genetic Algorithms (GA) and use of various Big data mechanisms over parallel GA models.

Keywords Parallel genetic algorithms · Big data · Map-Reduce · PGA models

6.1 Genetic Algorithms

Genetic Algorithms (GA) represent a group of evolutionary computing techniques that are used to search large spaces for discovering optimal solutions. They are heuristic techniques based on the principle of natural selection that perform random exploration in vast solution domains. They are currently used in a multitude of application areas such as engineering, business, and scientific applications. Being non-deterministic and stochastic in nature, they do not guarantee exact and same solutions each time they run. The basic principle of GA is based on biological

S. Sagnika (✉) · B.S.P. Mishra
School of Computer Engineering, KIIT University, Bhubaneswar, Odisha, India
e-mail: santwana.sagnika@gmail.com

B.S.P. Mishra
e-mail: mishra.bsp@gmail.com

S. Dehuri
Department of Information and Communication Technology,
Fakir Mohan University, Balasore, Odisha, India
e-mail: satchi.lapa@gmail.com

© Springer International Publishing Switzerland 2016
B.S.P. Mishra et al. (eds.), *Techniques and Environments for Big Data Analysis*,
Studies in Big Data 17, DOI 10.1007/978-3-319-27520-8_6

evolution, which is represented by the steps of selection, crossover, and mutation performed over a set of chromosomes, where each chromosome represents a particular solution to a problem [2, 9].

6.2 Parallel Genetic Algorithms

GAs are mostly applied to large problems involving complex datasets. Besides, the inherent features of GA include multiple iterations running on randomly generated data. These factors lead to a very high execution time. To reduce the running time and render the use of GA to more number of application areas, parallelization can prove to be a suitable mechanism. Since most part of GA consists of repetitive actions over different data, it is perfectly suited for parallelization. This has given rise to research into Parallel Genetic Algorithms (PGAs). PGAs are easier to implement, and more importantly, they satisfy the basic aim of parallelization, i.e., to provide a substantially better solution than the aggregate of solutions provided by its individual parts. For designing a PGA, either a single population can be used, or the entire population can be sub-divided into smaller populations and worked upon. The communication cost among the parallel systems also needs to be taken into account. Considering such factors, available research suggests high memory and CPU availability, better diversity, higher efficiency and provision of multiple solutions attained by implementation of PGAs. PGAs can also be applied for multi-objective optimization problems, wherein the parallel systems can seek different non-dominated solutions simultaneously [1, 2, 5]. Here the most popular PGA models are discussed.

PGAs can be broadly classified into three categories:
(1) Single-population Master-Slave GA,
(2) Fine-grained GA, and
(3) Coarse-grained GA.

Figure 6.1 shows an overview of the classification techniques.

Master-slave model is also known as global parallel GA. In this model, the solutions are represented in the form of a population which is shared globally, that is managed by a master processor. The other slave processors take the charge of

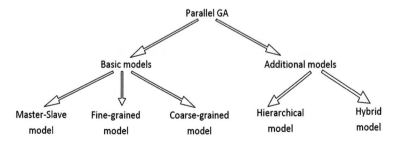

Fig. 6.1 Classification of PGA models

evaluating the fitness of the global population. The processes of selection and crossover occur globally. Since fitness evaluation is independent, communication occurs only when the slaves receive the population from the master and when they return the evaluated fitness value. Hence the overhead associated with communication is reduced (Fig. 6.2). When the algorithm pauses for all slaves to finish fitness evaluation before aggregating them and moving to the next iteration, it is known as synchronous. This kind of model increases the scale, calculation power and speed of GA. One such model is the Golub and Jakobovic model, where a shared memory is used to implement the master-slave concept (Fig. 6.3). This model suffers from getting trapped in local minima, hence a modification to this model has also been proposed. The modified model has a larger idle time (Fig. 6.4). To overcome these issues, a new model known as the Trigger model is designed. An efficient master-slave model depends on proper load balancing mechanisms.

The fine-grained model divides the entire population into a large number of smaller demes. It works on a single population. The population is spatially distributed. Every deme is restricted to be able to interact with only its immediate neighbors. Overlapping of the demes is permitted for the exchange of better solutions with the neighbors. Selection and crossover occur within the demes, or at most within the restricted neighborhood. Such a model is suitable for systems which have a high degree of parallelism. This model is also known as massively parallel GA (mpGA) or cellular GA (cGA). The increase in population size and neighborhood size can

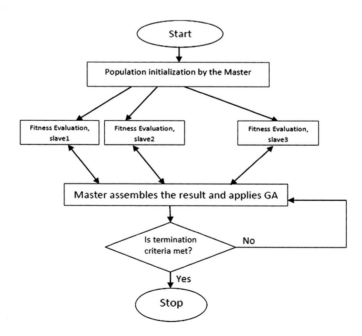

Fig. 6.2 Global parallelization model

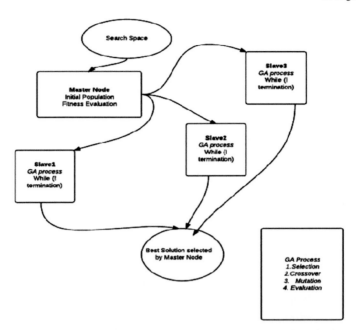

Fig. 6.3 Golub and Jakobovic model

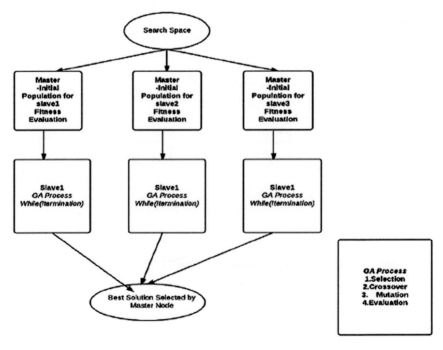

Fig. 6.4 Golub and Jakobovic modified model

Fig. 6.5 Coarse-grained
model

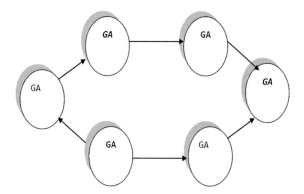

adversely affect the performance of the algorithm. Manderick et al. have proposed
an implementation of an mpGA using a 2-D grid distribution.

The coarse-grained model is also known as distributed GA (dGA) or island
GA. It consists of a set of sub-populations which perform migration, i.e., exchange
of individuals, occasionally. These methods are highly popular, and involve a set
of sub-populations that evolve in parallel. They perform migration, by exchange of
chromosomes between processes. Important factors to consider are the number and
size of demes and the topology of inter-connections. Grosso has done path-breaking
work in this field, discovering that a threshold migration rate has to be taken care
of, so that the performance is not degraded. Further work was done by Tanese who
implemented a 4-D hypercube model and explored the various factors that influence
migration. Designing such a multiple-deme parallel GA involves complex decisions
and choices (Fig. 6.5).

In addition, a combination of fine and coarse-grained GAs can also be employed
to realize parallelization. One such approach is to perform global parallelization
on each deme of a coarse-grained system. This leads to good speed-up at same
complexity [6].

6.3   Parallel Genetic Algorithms in Big Data Analysis

6.3.1   Map Reduce

Map Reduce is a Google framework for executing functions in batch mode that can
operate on a huge amount of data. It is followed by the Hadoop software since it
provides a high level of scalability. There are two major tasks in the process. The
first is Map, where the input data is broken down into smaller parts (key-value pairs)
which is suitable for further processing. The second is Reduce, which takes the above
output and combines them into new key-value pairs. The framework facilitates the
parallel processing of large datasets in cluster and grid systems [3, 4, 10].

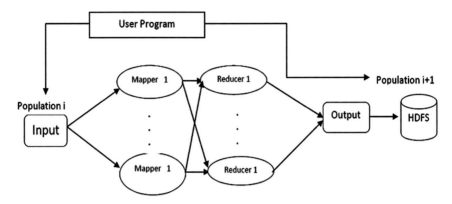

Fig. 6.6 Coarse grained parallelization using map reduce

PGA can be used in the Map and Reduce processes. Here Map Reduce using various PGA models is discussed.

6.3.1.1 Global Parallelization Using Map Reduce

Global Parallelization follows the Master-Slave model where one system becomes the master and the rest become slaves. The master node performs and stores most of the operations. It sends individuals to the slaves who perform selection operation on them and return the best selected individuals to the master. The main program is in charge of creation and initialization of the population and start the process. It also checks the termination criteria and determines when to stop the execution. The following algorithm shows the steps of the process.

Algorithm 1

Map(last_gen, next_gen)
{
curr_gen=0;
pop=CreateRandomPopulation(pop_size);
Fitness_function(pop);
while (curr_gen<last_gen) do
pop=Next_gen(pop);
curr_gen=curr_gen+1;
end while
opt=best from Fitness_function(pop);
return opt;
}
Next_gen(pop)
{

```
for (index in pop) do
parents=Selection(pop);
offspring=Crossover(parents);
offspring=Mutation(offspring);
add offspring to newOffspringSet;
end for
}
Reduce (opt)
{
for (index in opt) do
Check_opt(index);
end for
}
```

6.3.1.2   Coarse Grained Parallelization Using Map Reduce

The various computing nodes, termed as islands, further sub-divide the population among themselves and the genetic algorithm tasks are carried out by each node on its own group of individuals. Migration is carried out for exchanging the chromosomes among the nodes. A partition strategy needs to be defined so that every island of chromosomes is mapped to a Reducer. The Mapper randomly changes the islands, hence implementing migration, which results in chromosomes being assigned to a different Reducer (Fig. 6.6). The following algorithm shows the steps of the process.

Algorithm 2

```
Map(indiv_val,default_val)
{
Mapper(key, value);
pop=S1[0], S2[0],…, Sn[0] = indiv_val;
pop1=Evaluate(S1[0],S2[0],…,SN[0]);
Emit(default_val, pop1);
}
Reduce1 (indiv_val, key, val_list)
{
Reducer(key,val_list);
index=0;
for (each value in val_list) do
pop[index]=S1[index],S2[index],…,Sn[index]=indiv_val;
index++;
end for
pop1=Selection(pop);
for (each value in pop1) do
Emit(indiv_val,1);
}
Reduce2(key,value,indiv_key)
```

```
{
Find_Reduce(key,value);
pop=S1[0],S2[0],...,Sn[0]=indiv_key;
Emit(pop,1);
}
Coordinator()
{
MapReduce();
index=0;
pop[0]=S1[0],S2[0],...,Sn[0];
Evaluate(S1[0],S2[0],...,Sn[0]);
while(T[pop[index]]==false) do
pop1[index]=Crossover(pop[index]);
pop1[index]=Mutation(pop[index]);
Send_to_scheduler(pop1(index));
pop[index+1]=Recv_from_scheduler(index);
T=index+1;
return (pop[index]);
}
```

6.3.1.3 Fine Grained Parallelization Using Map Reduce

A slight variation in the previous proposal can lead to the implementation of Map Reduce over a fine-grained model. Each Mapper takes a single chromosome which is then passed to a randomly-selected Reducer. A pseudo-random function is used to generate the neighborhood as required by the model.

6.3.2 Compute Unified Device Architecture (CUDA)

It is a parallel computing model developed by NVIDIA, to perform parallel computations on Graphics Processing Units (GPUs). CUDA can take the advantage of parallelism by running on both GPUs and CPUs. In this architecture, programmers can access the GPU memory, and hence, make beneficial use of parallelization. A CUDA program execution is done at two locations- the host (CPU) and the device (GPU). CUDA is based on the C language. It can generate functions known as kernels. These kernels run in parallel on different threads during execution. Batches of threads can be grouped into blocks. The program is hosted on the CPU and the computations are done on the GPU who returns the results to the CPU.

6.3.2.1 Master-Slave Parallelization Using CUDA

The master-slave model does not require anything specific about the underlying architecture, so it can be implemented on any platform. The tasks are delegated to the GPU whose memory is extensively used. This reduces the communication between devices. The transfers that occur require lesser bandwidth. The algorithm is further modified to improve the performance and have a better distinction between the CPU that acts as a master node, controlling the process and issuing commands for operations that are to be carried out by the GPU or slave nodes. The following algorithm shows the steps that are followed [7–9, 11, 12].

Algorithm 3

Initialize the system
Set sizes of thread blocks and grid for maximum parallelization
***on GPU:** {Create and initialize the population}*
while TRUE do
on GPU:
{
Assess fitness of individuals
Assess statistics and state
Check for termination condition
}
if TRUE then
end algorithm;
end while;
else
continue
end if
on GPU:
{
Choose best individuals
Perform elitism
Perform scaling
Perform selection
Perform crossover
Perform mutation
Perform replacement to create new generation
}
end while
Provide results

6.3.2.2   Coarse-Grained Parallelization Using CUDA

An island-based GA can be taken, implementing migration in a single-directional ring topology. The CPU parallelizes the initial generation and sends them to the GPUs through the main memory. Then the kernels are created and the population is divided into input threads. CUDA threads control individuals stored on shared on-chip memory. The steps of the GA take place in parallel. CUDA block barriers separate the operations for better consistency. Migration is done asynchronously. After termination condition is reached, the threads write back the results to the main memory.

6.3.2.3   Fine-Grained Parallelization Using CUDA

For implementing fine-grained parallelization, a 2-dimensional matrix can be taken for representing the threads. This leads to simultaneous access of all genes of a chromosome. The user can provide a fitness evaluation kernel that utilizes the resources efficiently. This provides optimal performance to the user. This approach is highly useful for intensive calculations (Fig. 6.7).

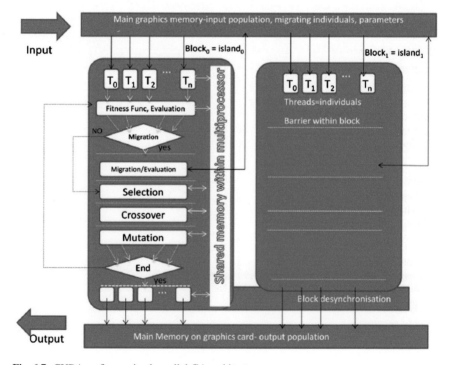

Fig. 6.7 CUDA on fine-grained parallel GA architecture

6.4 Summary

For most computational needs, it is highly necessary to reach the solution in the minimum possible time. Parallelization provides an efficient way to reduce the running time and make optimal use of resources available. Applying parallelization on GAs prove to be extremely effective since the long execution time is drastically reduced. This chapter has discussed the various PGA models, along with the application of Big data tools to implement the various PGA models, which have an extensive applicability in problems involving large and complex datasets. As future work, more Big data tools can be applied on PGA models with efficient algorithms that can fully exploit the parallelization and attain maximum speedup.

References

1. Alba, E., & Troya, J. M. (1999). A survey of parallel distributed genetic algorithms. *Complexity*, 4(4), 31–52.
2. Cantu-Paz, E. (1998). A survey of parallel genetic algorithms. *Calculateurs Paralleles, Reseaux Et Systems Repartis*, *10*,
3. Geronimo, L. D., Ferrucci, F., Murolo, A., & Sarro, F. (2012). A parallel genetic algorithm based on hadoop mapreduce for the automatic generation of junit test suites. In *IEEE Fifth International Conference on Software Testing, Verification and Validation*, pp. 785–793.
4. Golub, M., & Budin, L. (2000). An asynchronous model of global parallel genetic algorithms. In *Proceedings of the 22nd International Conference on Information Technology Interfaces*, pp. 363–368.
5. Hassani, A., & Treijs, J. (1975). An overview of standard and parallel genetic algorithms.
6. Mishra, B. S. P., Dehuri, S., Mall, R., & Ghosh, A. (2011). Parallel single and multiple objectives genetic algorithms: A survey. *International Journal of Applied Evolutionary Computation*, 2(2), 21–58.
7. Mivule, K., Harvey, B., Cobb, C., & Sayed, H. E. (2014). A review of cuda, mapreduce, and pthreads parallel computing models. *CoRR*, pp. 1–10.
8. Oiso, M., Matsumura, Y., Yasuda, T., & Ohkura, K. (2011). Implementing genetic algorithms to Cuda environment using data parallelization. *Tehnicki vjesnik/Technical Gazette*, *18*(4), 511–517.
9. Pospichal, P., Jaros, J., & Schwarz, J. (2010). Parallel genetic algorithm on the CUDA architecture. In *EvoApplications*, pp. 442–251.
10. Rahate, K. S., & Lobo, L. M. R. J. (2013). A novel technique for parallelization of genetic algorithm using Hadoop. *International Journal of Engineering Trends and Technology (IJETT)*, 4(8), 3328–3331.
11. Shah, R., Narayanan, P. J., & Kothapalli, K. (2010). GPU-accelerated genetic algorithms. In *Proceedings of the Workshop on Parallel Architectures for Bio-inspired Algorithms*, pp. 27–34.
12. Sharma, M., & Soni, P. (2014). Comparative study of parallel programming models to compute complex algorithm. *International Journal of Computer Applications*, 96(19), 9–12.

Chapter 7
Evolutionary Algorithm Based Techniques to Handle Big Data

Ghosh Sanchita and Desarkar Anindita

Abstract Big data is an all-encompassing term for any collection of data sets so large and complex that it becomes difficult to process them using traditional data processing applications. So Data mining, which has as a goal to extract knowledge from large databases, has become a challenge for this large and complex data set [52]. Other challenges in handling Big data include analysis, capture, curation, search, sharing, storage, transfer, visualization, and privacy violations. To extract and handle this large amount of knowledge, a database may be considered as a large search space, and a mining algorithm as a search strategy. In general, a search space consists of an enormous number of elements, making an exhaustive search infeasible. Therefore, efficient search strategies are of vital importance. Search strategies based on Evolutionary algorithms have been applied successfully in a wide range of applications. In this chapter, we discuss about Big data and limitations of standard algorithms handling them, Evolutionary Algorithm and their advantages in handling Big data, Commonly used Evolutionary algorithm—Genetic Algorithm and the various application areas where genetic Algorithm plays evolutionary role in the large and complex search space.

Keywords Genetic algorithm · Big data · Data acquisition and recording · Information extraction and cleaning · Interpretation · Heterogeneity and incompleteness · Representation · Evaluation function · Population · Parent selection mechanism · Variation operators · Survivor selection mechanism · Initialization · Termination condition · Hybrid selection method · Enhanced roulette wheel selection · Tournament selection · Selection operator

G. Sanchita (✉)
Birla Institute of Technology, Mesra, Ext. Center, Kolkata, West Bengal, India
e-mail: sanchitag@bitmesra.ac.in

D. Anindita
Cognizant Technology Solutions, Salt Lake, Kolkata 700091, West Bengal, India

© Springer International Publishing Switzerland 2016
B.S.P. Mishra et al. (eds.), *Techniques and Environments for Big Data Analysis*, Studies in Big Data 17, DOI 10.1007/978-3-319-27520-8_7

7.1 Introduction

The amount of data stored in databases continues to grow fast. Intuitively, this large amount of stored data contains valuable hidden knowledge, which could be used to improve the decision-making process of an organization. For instance, data about previous sales might contain interesting relationships between products and customers. The discovery of such relationships can be very useful to increase the sales of a company. However, the number of human data analysts grows at a much smaller rate than the amount of stored data. Thus, there is a clear need for efficient search strategies for extracting knowledge from data.

In this scenario, **Evolutionary algorithms** can be used as a feasible solution because of their attractive features that enable them to resolve some of the drawbacks faced in handling Big data with conventional techniques and enable them to discover novel solutions, such as their robustness when dealing with noisy data, and their ability to interpret data without any a priori knowledge.

Also in the **Big data** environment, finding a solution in a very complex search space or exhaustive search is a big challenge due to it's enormous volume, velocity and variety. The traditional search techniques are too slow for finding an acceptable solution where the Evolutionary algorithms are extremely useful as it requires little information to search effectively in this kind of large/complex search space. For exhaustive search, it does not require searching each and every point in the search space, exploration is the process of visiting entirely new regions of a search space. This is controlled by Crossover and Mutation operators by creating diversity in the population. On the other hand, exploitation which is the reduction of diversity by focusing on the better individuals is taken care by the selection process. As a result, a good search technique must find a good balance between these two important factors: exploration and exploitation to find a global optimum. A trade off ensures that good solutions go to the next generation more frequently than poor solutions.

In this chapter, we have discussed about Big data and the limitations of standard algorithms in handling them, Evolutionary Algorithm and their advantages in handling Big data, Evolutionary algorithm and it's major components along with their brief description, the major benefits of Evolutionary algorithm in Big data analysis, Criteria for using Evolutionary algorithm, Brief description of the Commonly used Evolutionary algorithm—Genetic Algorithm and it's various selection operators, the various application areas where genetic Algorithm plays crucial role in the Big data environment.

The rest of the chapter is organized as follows. Section 7.2 describes the basic concept of Big data. Limitations of standard algorithm in handling Big data is included in Sect. 7.3. Basic description of Evolutionary algorithm and its major components are provided in Sect. 7.4. Section 7.5 highlights the benefit of Evolutionary algorithm in the analysis of Big data; Sect. 7.6 describes the situations when the Evolutionary algorithm is useful. The common variations of Evolutionary algorithm are described in Sects. 7.7 and 7.8 gives the brief overview of the applications where Genetic algorithm, a branch of Evolutionary algorithm can be used as a good solution technique and the conclusions are summarized in Sect. 7.8.

7.2 What Is Big Data

Big data usually includes data sets with sizes beyond the ability of commonly used software tools to capture, curate, manage, and process data within a tolerable elapsed time. Big data size is a constantly moving target, as of 2012 ranging from a few dozen terabytes to many petabytes of data. Big data needs a set of techniques and technologies that require new forms of integration to uncover large hidden values from large datasets that are diverse, complex, and of a massive scale.

In a 2001 research report and related lectures, META Group (now Gartner) analyst Doug Laney defined data growth challenges and opportunities as being three-dimensional, i.e. increasing volume (amount of data), velocity (speed of data in and out), and variety (range of data types and sources). Gartner, and nosw much of the industry, continue to use this 3Vs model for describing Big data. In 2012, Gartner updated its definition as follows: Big data is **high volume, high velocity, and/or high variety information** assets that require new forms of processing to enable enhanced decision making, insight discovery and process optimization. Additionally, a new V **Veracity** is added by some organizations to describe it [2].

If Gartner's definition (the 3Vs) is still widely used, the growing maturity of the concept fosters a more sound difference between Big data and Business Intelligence, regarding data and their use:

Business Intelligence uses descriptive statistics with data with high information density to measure things, detect trends etc.

Big data uses inductive statistics and concepts from nonlinear system identification to infer laws (regressions, nonlinear relationships, and causal effects) from large sets of data with low information density to reveal relationships, dependencies and perform predictions of outcomes and behaviors.

Big data can also be defined as "Big data is a large volume unstructured data which cannot be handled by standard database management systems like DBMS, RDBMS or ORDBMS".

Big data is an all-encompassing term for any collection of data sets so large and complex that it becomes difficult to process them using traditional data processing applications.

The challenges include analysis, capture, curation, search, sharing, storage, transfer, visualization, and privacy violations. The trend to larger data sets is due to the additional information derivable from analysis of a single large set of related data, as compared to separate smaller sets with the same total amount of data, allowing correlations to be found to "spot business trends, prevent diseases, combat crime and so on."

Scientists regularly encounter limitations due to large data sets in many areas, including meteorology, genomics, complex physics simulations, and biological and environmental research. The limitations also affect Internet search, finance and business informatics. Data sets grow in size in part because they are increasingly being gathered by ubiquitous information-sensing mobile devices, aerial sensory technologies (remote sensing), software logs, cameras, microphones, radio-frequency identification (RFID) readers, and wireless sensor networks. The world's technological per-capita capacity to store information has roughly doubled every 40 months since the 1980s; as of 2012, every day 2.5 EB (2.51018) of data were created; as of 2014[update], every day 2.3 ZB (2.31021) of data were created. The challenge for large enterprises is determining who should own Big data initiatives that straddle the entire organization.

Big data is difficult to work with using most relational database management systems and desktop statistics and visualization packages, requiring instead "massively parallel software running on tens, hundreds, or even thousands of servers". What is considered "Big data" varies depending on the capabilities of the organization managing the set, and on the capabilities of the applications that are traditionally used to process and analyze the data set in its domain. Big data is a moving target; what is considered to be "Big" today will not be so years ahead. "For some organizations, facing hundreds of gigabytes of data for the first time may trigger a need to reconsider data management options. For others, it may take tens or hundreds of terabytes before data size becomes a significant consideration."

7.3 Limitations of Standard Algorithms in Handling Big Data

The promise of data-driven decision-making is now being recognized broadly, and there is growing enthusiasm for the notion of "Big data". While the promise of Big data is real—for example, it is estimated that Google alone contributed 54 billion dollars to the US economy in 2009—there is currently a wide gap between its potential and its realization [2].

Heterogeneity, scale, timeliness, complexity, and privacy problems with Big data impede progress at all phases of the pipeline that can create value from data. Much data today is not natively in structured format; for example, tweets and blogs are weakly structured pieces of text, while images and video are structured for storage and display, but not for semantic content and search: transforming such content into a structured format for later analysis is a major challenge. The value of data explodes when it can be linked with other data, thus data integration is a major creator of value. Since most data is directly generated in digital format today, we have the opportunity and the challenge both to influence the creation to facilitate later linkage and to automatically link previously created data. Data analysis, organization, retrieval, and modeling are other foundational challenges. Data analysis is a clear bottleneck in many applications, both due to lack of scalability of the underlying algorithms and due to the complexity of the data that needs to be analyzed. Finally, presentation of the results and its interpretation by non-technical domain experts is crucial to extracting actionable knowledge [2].

We are awash in a flood of data today. In a broad range of application areas, data is being collected at unprecedented scale. Decisions that previously were based on guesswork, or on painstakingly constructed models of reality, can now be made based on the data itself. Such Big data analysis now drives nearly every aspect of our modern society, including mobile services, retail, manufacturing, financial services, life sciences, and physical sciences.

Scientific research has been revolutionized by Big data. The Sloan Digital Sky Survey has today become a central resource for astronomers the world over. The field of Astronomy is being transformed from one where taking pictures of the sky was a large part of an astronomers job to one where the pictures are all in a database already and the astronomers task is to find interesting objects and phenomena in the database. In the biological sciences, there is now a well-established tradition of depositing scientific data into a public repository, and also of creating public databases for use by other scientists. In fact, there is an entire discipline of bioinformatics that is largely devoted to the curation and analysis of such data. As technology advances, particularly with the advent of Next Generation Sequencing, the size and number of experimental data sets available is increasing exponentially.

While the potential benefits of Big data are real and significant, and some initial successes have already been achieved (such as the Sloan Digital Sky Survey), there remain many technical challenges that must be addressed to fully realize this potential. The sheer size of the data, of course, is a major challenge, and is the one that is most easily recognized. However, there are others. Industry analysis companies like to point out that there are challenges not just in Volume, but also in Variety and Velocity, and that companies should not focus on just the first of these. By Variety, they usually mean heterogeneity of data types, representation, and semantic interpretation. By Velocity, they mean both the rate at which data arrive and the time in which it must be acted upon. While these three are important, this short list fails to include additional important requirements such as privacy and usability.

The analysis of Big data involves multiple distinct phases as shown in the Fig. 7.1 below, each of which introduces challenges. Many people unfortunately focus just on the analysis/modeling phase: while that phase is crucial, it is of little use without the other phases of the data analysis pipeline. Even in the analysis phase, which has received much attention, there are poorly understood complexities in the context of multi-tenanted clusters where several users programs run concurrently. Many significant challenges extend beyond the analysis phase. For example, Big data has to be managed in context, which may be noisy, heterogeneous and not include an upfront model. Doing so raises the need to track provenance and to handle uncertainty and error: topics that are crucial to success, and yet rarely mentioned in the same breath as Big data. Similarly, the questions to the data analysis pipeline will typically not all be laid out in advance. We may need to figure out good questions based on the data. Doing this will require smarter systems and also better support for user interaction with the analysis pipeline. In fact, we currently have a major bottleneck in the number of people empowered to ask questions of the data and analyze it. We can drastically increase this number by supporting many levels of engagement with the data, not all requiring deep database expertise. Solutions to problems such as this will not come from incremental improvements to business as usual such as industry may make on its own. Rather, they require us to fundamentally rethink how we manage data analysis.

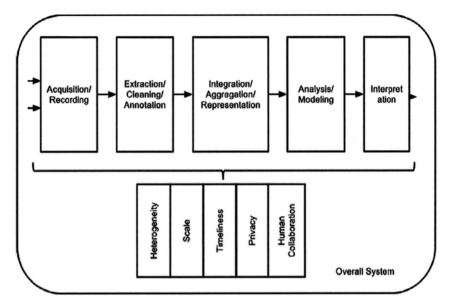

Fig. 7.1 The Big data analysis pipeline. Major steps in analysis Big data are shown in the flow at *top*. Below it are Big data needs that make these tasks challenging

Fortunately, existing computational techniques can be applied, either as is or with some extensions, to at least some aspects of the Big data problem. For example, relational databases rely on the notion of logical data independence: users can think about what they want to compute, while the system (with skilled engineers designing those systems) determines how to compute it efficiently. Similarly, the SQL standard and the relational data model provide a uniform, powerful language to express many query needs and, in principle, allows customers to choose between vendors, increasing competition. The challenge ahead of us is to combine these healthy features of prior systems as we devise novel solutions to the many new challenges of Big data.

7.3.1 Challenges in Data Acquisition and Recording Phase

Big data does not arise out of a vacuum: it is recorded from some data generating source. For example, consider our ability to sense and observe the world around us, from the heart rate of an elderly citizen, and presence of toxins in the air we breathe, to the planned square kilometer array telescope, which will produce up to 1 million terabytes of raw data per day. Similarly, scientific experiments and simulations can easily produce petabytes of data today.

Much of this data is of no interest, and it can be filtered and compressed by orders of magnitude. One challenge is to define these filters in such a way that they do not discard useful information. For example, suppose one sensor reading differs substantially from the rest: it is likely to be due to the sensor being faulty, but how can we be sure that it is not an artifact that deserves attention? In addition, the data collected by these sensors most often are spatially and temporally correlated (e.g., traffic sensors on the same road segment). We need research in the science of data reduction that can intelligently process this raw data to a size that its users can handle while not missing the needle in the haystack. Furthermore, we require on-line analysis techniques that can process such streaming data on the fly, since we cannot afford to store first and reduce afterward.

The second big challenge is to automatically generate the right metadata to describe what data is recorded and how it is recorded and measured. For example, in scientific experiments, considerable detail regarding specific experimental conditions and procedures may be required to be able to interpret the results correctly, and it is important that such metadata be recorded with observational data. Metadata acquisition systems can minimize the human burden in recording metadata. Another important issue here is data provenance. Recording information about the data at its birth is not useful unless this information can be interpreted and carried along through the data analysis pipeline. For example, a processing error at one step can render subsequent analysis useless; with suitable provenance, we can easily identify all subsequent processing that dependent on this step. Thus we need research both into generating suitable metadata and into data systems that carry the provenance of data and its metadata through data analysis pipelines [2].

7.3.2 Challenges in Information Extraction and Cleaning Phase

Frequently, the information collected will not be in a format ready for analysis. For example, consider the collection of electronic health records in a hospital, comprising transcribed dictations from several physicians, structured data from sensors and measurements (possibly with some associated uncertainty), and image data such as x-rays. We cannot leave the data in this form and still effectively analyze it. Rather we require an information extraction process that pulls out the required information from the underlying sources and expresses it in a structured form suitable for analysis. Doing this correctly and completely is a continuing technical challenge. Note that this data also includes images and will in the future include video; such extraction is often highly application dependent (e.g., what you want to pull out of an MRI is very different from what you would pull out of a picture of the stars, or a surveillance photo). In addition, due to the ubiquity of surveillance cameras and popularity of GPS-enabled mobile phones, cameras, and other portable devices, rich and high fidelity location and trajectory (i.e., movement in space) data can also be extracted.

We are used to thinking of Big data as always telling us the truth, but this is actually far from reality. For example, patients may choose to hide risky behavior and caregivers may sometimes mis-diagnose a condition; patients may also inaccurately recall the name of a drug or even that they ever took it, leading to missing information in (the history portion of) their medical record. Existing work on data cleaning assumes well-recognized constraints on valid data or well-understood error models; for many emerging Big data domains these do not exist [2].

7.3.3 Challenges in Data Integration, Aggregation, and Representation

Given the heterogeneity of the flood of data, it is not enough merely to record it and throw it into a repository. Consider, for example, data from a range of scientific experiments. If we just have a bunch of data sets in a repository, it is unlikely anyone will ever be able to find, let alone reuse, any of this data. With adequate metadata, there is some hope, but even so, challenges will remain due to differences in experimental details and in data record structure.

Data analysis is considerably more challenging than simply locating, identifying, understanding, and citing data. For effective large-scale analysis all of this has to happen in a completely automated manner. This requires differences in data structure and semantics to be expressed in forms that are computer understandable, and then robotically resolvable. There is a strong body of work in data integration that can provide some of the answers. However, considerable additional work is required to achieve automated error-free difference resolution.

Even for simpler analyses that depend on only one data set, there remains an important question of suitable database design. Usually, there will be many alternative ways in which to store the same information. Certain designs will have advantages over others for certain purposes, and possibly drawbacks for other purposes. Witness, for instance, the tremendous variety in the structure of bioinformatics databases with information regarding substantially similar entities, such as genes. Database design is today an art, and is carefully executed in the enterprise context by highly-paid professionals. We must enable other professionals, such as domain scientists, to create effective database designs, either through devising tools to assist them in the design process or through forgoing the design process completely and developing techniques so that databases can be used effectively in the absence of intelligent database design [2].

7.3.4 Challenges in Query Processing, Data Modeling and Analysis

Methods for querying and mining Big data are fundamentally different from traditional statistical analysis on small samples. Big data is often noisy, dynamic, heterogeneous, inter-related and untrustworthy. Nevertheless, even noisy Big data could be more valuable than tiny samples because general statistics obtained from frequent patterns and correlation analysis usually overpower individual fluctuations and often disclose more reliable hidden patterns and knowledge. Further, interconnected Big data forms large heterogeneous information networks, with which information redundancy can be explored to compensate for missing data, to crosscheck conflicting cases, to validate trustworthy relationships, to disclose inherent clusters, and to uncover hidden relationships and models.

Mining requires integrated, cleaned, trustworthy, and efficiently accessible data, declarative query and mining interfaces, scalable mining algorithms, and Big data computing environments. At the same time, data mining itself can also be used to help improve the quality and trustworthiness of the data, understand its semantics, and provide intelligent querying functions. As noted previously, real-life medical records have errors, are heterogeneous, and frequently are distributed across multiple systems. The value of Big data analysis in health care, to take just one example application domain, can only be realized if it can be applied robustly under these difficult conditions. On the flip side, knowledge developed from data can help in correcting errors and removing ambiguity. For example, a physician may write DVT as the diagnosis for a patient. This abbreviation is commonly used for both deep vein thrombosis and diverticulitis, two very different medical conditions. A knowledge-base constructed from related data can use associated symptoms or medications to determine which of two the physician meant.

Big data is also enabling the next generation of interactive data analysis with real-time answers. In the future, queries towards Big data will be automatically generated for content creation on websites, to populate hot-lists or recommendations, and to provide an ad hoc analysis of the value of a data set to decide whether to store or to discard it. Scaling complex query processing techniques to terabytes while enabling interactive response times is a major open research problem today.

A problem with current Big data analysis is the lack of coordination between database systems, which host the data and provide SQL querying, with analytics packages that perform various forms of non-SQL processing, such as data mining and statistical analyses. Todays analysts are impeded by a tedious process of exporting data from the database, performing a non-SQL process and bringing the data back. This is an obstacle to carrying over the interactive elegance of the first generation of SQL-driven OLAP systems into the data mining type of analysis that is in increasing demand. A tight coupling between declarative query languages and the functions of such packages will benefit both expressiveness and performance of the analysis [2].

7.3.5 Challenges in Interpretation

Having the ability to analyze Big data is of limited value if users cannot understand the analysis. Ultimately, a decision-maker, provided with the result of analysis, has to interpret these results. This interpretation cannot happen in a vacuum. Usually, it involves examining all the assumptions made and retracing the analysis. Furthermore, as we saw above, there are many possible sources of error: computer systems can have bugs, models almost always have assumptions, and results can be based on erroneous data. For all of these reasons, no responsible user will cede authority to the computer system. Rather she will try to understand, and verify, the results produced by the computer. The computer system must make it easy for her to do so. This is particularly a challenge with Big data due to its complexity. There are often crucial assumptions behind the data recorded. Analytical pipelines can often involve multiple steps, again with assumptions built in. The recent mortgage-related shock to the financial system dramatically underscored the need for such decision-maker diligence—rather than accept the stated solvency of a financial institution at face value, a decision-maker has to examine critically the many assumptions at multiple stages of analysis.

In short, it is rarely enough to provide just the results. Rather, one must provide supplementary information that explains how each result was derived, and based upon precisely what inputs. Such supplementary information is called the provenance of the (result) data. By studying how best to capture, store, and query provenance, in conjunction with techniques to capture adequate metadata, we can create an infrastructure to provide users with the ability both to interpret analytical results obtained and to repeat the analysis with different assumptions, parameters, or data sets.

Systems with a rich palette of visualizations become important in conveying to the users the results of the queries in a way that is best understood in the particular domain.

Furthermore, with a few clicks the user should be able to drill down into each piece of data that she sees and understand its provenance, which is a key feature to understanding the data. That is, users need to be able to see not just the results, but also understand why they are seeing those results. However, raw provenance, particularly regarding the phases in the analytics pipeline, is likely to be too technical for many users to grasp completely. One alternative is to enable the users to play with the steps in the analysis make small changes to the pipeline, for example, or modify values for some parameters. The users can then view the results of these incremental changes. By these means, users can develop an intuitive feeling for the analysis and also verify that it performs as expected in corner cases. Accomplishing this requires the system to provide convenient facilities for the user to specify analyses [2].

There are also few more challenges in handling Big data(described below) except the above ones which are common to all the phases [2].

7.3.6 Heterogeneity and Incompleteness

When humans consume information, a great deal of heterogeneity is comfortably tolerated. In fact, the nuance and richness of natural language can provide valuable depth. However, machine analysis algorithms expect homogeneous data, and cannot understand nuance. In consequence, data must be carefully structured as a first step in (or prior to) data analysis. Consider, for example, a patient who has multiple medical procedures at a hospital. We could create one record per medical procedure or laboratory test, one record for the entire hospital stay, or one record for all life-time hospital interactions of this patient. With anything other than the first design, the number of medical procedures and lab tests per record would be different for each patient. The three design choices listed have successively less structure and, conversely, successively greater variety. Greater structure is likely to be required by many (traditional) data analysis systems. However, the less structured design is likely to be more effective for many purposes for example questions relating to disease progression over time will require an expensive join operation with the first two designs, but can be avoided with the latter. However, computer systems work most efficiently if they can store multiple items that are all identical in size and structure. Efficient representation, access, and analysis of semi-structured data require further work.

Consider an electronic health record database design that has fields for birth date, occupation, and blood type for each patient. What do we do if one or more of these pieces of information is not provided by a patient? Obviously, the health record is still placed in the database, but with the corresponding attribute values being set to NULL. A data analysis that looks to classify patients by, say, occupation, must take into account patients for which this information is not known. Worse, these patients with unknown occupations can be ignored in the analysis only if we have reason to believe that they are otherwise statistically similar to the patients with known occupation for the analysis performed. For example, if unemployed patients are more likely to hide their employment status, analysis results may be skewed in that it considers a more employed population mix than exists, and hence potentially one that has differences in occupation-related health-profiles.

Even after data cleaning and error correction, some incompleteness and some errors in data are likely to remain. This incompleteness and these errors must be managed during data analysis. Doing this correctly is a challenge. Recent work on managing probabilistic data suggests one way to make progress [2].

7.3.7 Scale

Of course, the first thing anyone thinks of with Big data is its size. After all, the word big is there in the very name. Managing large and rapidly increasing volumes of data has been a challenging issue for many decades. In the past, this challenge was mitigated by processors getting faster, following Moores law, to provide us with the resources needed to cope with increasing volumes of data. But, there is a fundamental shift underway now: data volume is scaling faster than compute resources, and CPU speeds are static.

First, over the last five years the processor technology has made a dramatic shift—rather than processors doubling their clock cycle frequency every 18–24 months, now, due to power constraints, clock speeds have largely stalled and processors are being built with increasing numbers of cores. In the past, large data processing systems had to worry about parallelism across nodes in a cluster; now, one has to deal with parallelism within a single node. Unfortunately, parallel data processing techniques that were applied in the past for processing data across nodes dont directly apply for intra-node parallelism, since the architecture looks very different; for example, there are many more hardware resources such as processor caches and processor memory channels that are shared across cores in a single node. Furthermore, the move towards packing multiple sockets (each with 10s of cores) adds another level of complexity for intra-node parallelism. Finally, with predictions of dark silicon, namely that power consideration will likely in the future prohibit us from using all of the hardware in the system continuously, data processing systems will likely have to actively manage the power consumption of the processor. These unprecedented changes require us to rethink how we design, build and operate data processing components.

The second dramatic shift that is underway is the move towards cloud computing, which now aggregates multiple disparate workloads with varying performance goals (e.g. interactive services demand that the data processing engine return back an answer within a fixed response time cap) into very large clusters. This level of sharing of resources on expensive and large clusters requires new ways of determining how to run and execute data processing jobs so that we can meet the goals of each workload cost-effectively, and to deal with system failures, which occur more frequently as we operate on larger and larger clusters (that are required to deal with the rapid growth in data volumes). This places a premium on declarative approaches to expressing programs, even those doing complex machine learning tasks, since global optimization across multiple users programs is necessary for good overall performance. Reliance on user-driven program optimizations is likely to lead to poor cluster utilization, since users are unaware of other users programs. System-driven holistic optimization requires programs to be sufficiently transparent, e.g., as in relational database systems, where declarative query languages are designed with this in mind.

A third dramatic shift that is underway is the transformative change of the traditional I/O subsystem. For many decades, hard disk drives (HDDs) were used to store persistent data. HDDs had far slower random IO performance than sequential IO performance, and data processing engines formatted their data and designed their query processing methods to work around this limitation. But, HDDs are increasingly being replaced by solid state drives today, and other technologies such as Phase Change Memory are around the corner. These newer storage technologies do not have the same large spread in performance between the sequential and random I/O performance, which requires a rethinking of how we design storage subsystems for data processing systems. Implications of this changing storage subsystem potentially touch every aspect of data processing, including query processing algorithms, query scheduling, database design, concurrency control methods and recovery methods [2].

7.3.8 Timeliness

The flip side of size is speed. The larger the data set to be processed, the longer it will take to analyze. The design of a system that effectively deals with size is likely also to result in a system that can process a given size of data set faster. However, it is not just this speed that is usually meant when one speaks of Velocity in the context of Big data.

There are many situations in which the result of the analysis is required immediately. For example, if a fraudulent credit card transaction is suspected, it should ideally be flagged before the transaction is completed potentially preventing the transaction from taking place at all. Obviously, a full analysis of a users purchase history is not likely to be feasible in real-time. Rather, we need to develop partial results in advance so that a small amount of incremental computation with new data can be used to arrive at a quick determination.

Given a large data set, it is often necessary to find elements in it that meet a specified criterion. In the course of data analysis, this sort of search is likely to occur repeatedly. Scanning the entire data set to find suitable elements is obviously impractical. Rather, index structures are created in advance to permit finding qualifying elements quickly. The problem is that each index structure is designed to support only some classes of criteria. With new analyses desired using Big data, there are new types of criteria specified, and a need to devise new index structures to support such criteria. For example, consider a traffic management system with information regarding thousands of vehicles and local hot spots on roadways. The system may need to predict potential congestion points along a route chosen by a user, and suggest alternatives. Doing so requires evaluating multiple spatial proximity queries working with the trajectories of moving objects. New index structures are required to support such queries. Designing such structures becomes particularly challenging when the data volume is growing rapidly and the queries have tight response time limits [2].

7.3.9 Privacy

The privacy of data is another huge concern, and one that increases in the context of Big data. For electronic health records, there are strict laws governing what can and cannot be done. For other data, regulations, particularly in the US, are less forceful. However, there is great public fear regarding the inappropriate use of personal data, particularly through linking of data from multiple sources. Managing privacy is effectively both a technical and a sociological problem, which must be addressed jointly from both perspectives to realize the promise of Big data.

Consider, for example, data gleaned from location-based services. These new architectures require a user to share his/her location with the service provider, result ing in obvious privacy concerns. Note that hiding the users identity alone without hiding her location would not properly address these privacy concerns. An attacker or a (potentially malicious) location-based server can infer the identity of the query source from its (subsequent) location information. For example, a users location information can be tracked through several stationary connection points (e.g., cell towers). After a while, the user leaves a trail of packet crumbs which could be associated to a certain residence or office location and thereby used to determine the users identity. Several other types of surprisingly private information such as health issues (e.g., presence in a cancer treatment center) or religious preferences (e.g., presence in a church) can also be revealed by just observing anonymous users movement and usage pattern over time. In general, Barabsi et al. showed that there is a close correlation between peoples identities and their movement patterns [27]. Note that hiding a user location is much more challenging than hiding his/her identity. This is because with location-based services, the location of the user is needed for a successful data access or data collection, while the identity of the user is not necessary.

There are many additional challenging research problems. For example, we do not know yet how to share private data while limiting disclosure and ensuring sufficient data utility in the shared data. The existing paradigm of differential privacy is a very important step in the right direction, but it unfortunately reduces information content too far in order to be useful in most practical cases. In addition, real data is not static but gets larger and changes over time; none of the prevailing techniques results in any useful content being released in this scenario. Yet another very important direction is to rethink security for information sharing in Big data use cases. Many online services today require us to share private information (think of Facebook applications), but beyond record-level access control we do not understand what it means to share data, how the shared data can be linked, and how to give users fine-grained control over this sharing [2].

7.3.10   Human Collaboration

In spite of the tremendous advances made in computational analysis, there remain many patterns that humans can easily detect but computer algorithms have a hard time finding. Indeed, CAPTCHAs exploit precisely this fact to tell human web users apart from computer programs. Ideally, analytics for Big data will not be all computational rather it will be designed explicitly to have a human in the loop. The new sub-field of visual analytics is attempting to do this, at least with respect to the modeling and analysis phase in the pipeline. There is similar value to human input at all stages of the analysis pipeline.

In today's complex world, it often takes multiple experts from different domains to really understand what is going on. A Big data analysis system must support input from multiple human experts, and shared exploration of results. These multiple experts may be separated in space and time when it is too expensive to assemble an entire team together in one room. The data system has to accept this distributed expert input, and support their collaboration.

A popular new method of harnessing human ingenuity to solve problems is through crowd-sourcing. Wikipedia, the online encyclopedia, is perhaps the best known example of crowd-sourced data. We are relying upon information provided by unvetted strangers. Most often, what they say is correct. However, we should expect there to be individuals who have other motives and abilities some may have a reason to provide false information in an intentional attempt to mislead. While most such errors will be detected and corrected by others in the crowd, we need technologies to facilitate this. We also need a framework to use in analysis of such crowd-sourced data with conflicting statements. As humans, we can look at reviews of a restaurant, some of which are positive and others critical, and come up with a summary assessment based on which we can decide whether to try eating there. We need computers to be able to do the equivalent. The issues of uncertainty and error become even more pronounced in a specific type of crowd-sourcing, termed participatory-sensing. In this case, every person with a mobile phone can act as a multi-modal sensor collecting various types of data instantaneously (e.g., picture, video, audio, location, time, speed, direction, acceleration). The extra challenge here is the inherent uncertainty of the data collection devices. The fact that collected data are probably spatially and temporally correlated can be exploited to better assess their correctness. When crowd-sourced data is obtained for hire, such as with Mechanical Turks, much of the data created may be with a primary objective of getting it done quickly rather than correctly. This is yet another error model, which must be planned for explicitly when it applies [2].

7.4 What Is Evolutionary Algorithm

Evolutionary Algorithms (EAs) mimic natural evolutionary principles to constitute search and optimization procedures.

Though there are many different variations of EAs, the common underlying idea behind all techniques is same: given a population of individuals the environmental pressure causes natural selection (survival of the fittest) and this causes a rise of fitness in the population. Given a quality function to be examined, we can randomly create a set of candidate solutions, i.e. elements of the functions domain and apply the quality function as an abstract fitness measure the higher the better. Based on this fitness, some of the better candidates are chosen to seed the next generation by applying recombination and/or mutation to them. Recombination is an operator applied to two or more selected candidates (the so called parents) and results one/more new candidates (the children). Mutation is applied on one candidate and results in one new candidate. Execution of Recombination and mutation leads to a set of new candidates (the offspring) that compete based on their fitness (and possibly age) with the old ones for a place in the next generation. This process can be iterated until a candidate with sufficient quality (a solution) is found or previously set computational limit is reached [20].

In this process, there are two fundamental forces that form the basis of evolutionary systems.

- Variation operators (recombination and mutation) create the necessary diversity and thereby facilitate novelty, while
- Selection acts as a force pushing quality

The combined application of variation and selection generally leads to improving fitness values in consecutive populations.

Many components of such as evolutionary process are stochastic. During selection, fitter individuals have a higher chance of selection to be selected than the less fit ones, but typically even the weak individuals have a chance to become a parent or to survive. For recombination of individuals, the choice of which pieces will be recombined is random. Similarly for mutation, the pieces that will be mutated within a candidate solution, and the new pieces replacing them, are chosen randomly. The general scheme of EA is given below in the flowchart and pseudo code fashion respectively [20].

Evolutionary Algorithm –Pseudo code Representation:
BEGIN
INITIALIZE population with random candidate solutions;
EVALUATE each candidate;
REPEAT UNTIL (TERMINATION CONDITION IS SATISFIED) DO
1. *SELECT parents;*
2. *RECOMBINE pairs of parents;*
3. *MUTATE the resulting offspring;*
4. *EVALUATE new candidates;*
5. *SELECT individuals for the next generation*
OD
END

EAs have a number of components, procedures or operators that must be specified in order to define a particular EA. The most important components are given below.

- **Representation (Definitions of individuals)**
- **Evaluation Function (or fitness function)**
- **Population**
- **Parent Selection mechanism**
- **Variation operators, Recombination and Mutation**
- **Survivor selection mechanism (Replacement)**

Also the initialization and termination conditions must be defined properly for correct running of an algorithm.

The above components are described below in a brief manner (Fig. 7.2).

Fig. 7.2 Evolutionary
algorithm-flowchart
representation

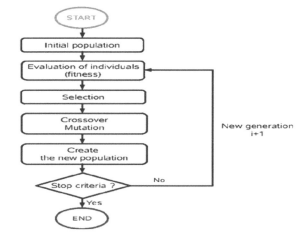

7.4.1 Representation (Definition of Individuals)

The 1st step in defining an EA is to link the real world to the EA world, that is to
set up a bridge between the original problem context and the problem solving space
where the evolution will take space. Objects forming possible solutions within the
original problem context are referred to as phenotypes, their encoding, the individuals within the EA are called genotypes. The first design step is commonly called
representation, as if amounts to specifying a mapping from the phenotypes on to
a set of genotypes that are said to represent these phenotypes. For instance given
an optimization problem on integers, the given set of integers would form the set of
phenotypes. Then one could decide to represent by their binary code, hence 18 would
be seen as a phenotype and 10010 as a genotype representation. It is important to
understand that the phenotype space can be very different from the genotype space
and that the whole evolutionary search takes place in the genotype space [20].

7.4.2 Evaluation Function (Fitness Function)

The role of the evaluation function is to represent the requirement to adapt to. It
forms the basis for selection and there by it facilitates improvements. More accu-
rately it defines what improvement means. From the problem solving perspective, it
represents the task to solve in the evolutionary context. Technically its a functional
or procedure that assigns a quality measure to genotypes. Typically this function is
composed from a quality measure in the phenotype space and the inverse representation. To remain with the above example, if we were to maximize x × x (x^2) on
integers, the fitness of the genotype 10010 could be defined as the square of its corresponding phenotype: $18 \times 18 = 324$. The evaluation function is commonly called
the fitness function in EA [20].

7.4.3 Population

The role of the population is to hold possible solutions. A population is a multiset of genotypes. The population forms the unit of evolution. Individuals are static objects not changing or adapting, it is the population that does. Given a representation defining a population can be as simple as specifying how many individuals are in it, that is setting the population size [20].

7.4.4 Parent Selection Mechanism

The role of parent selection or mating selection is to distinguish among individuals based on their quality, in particular, to allow the better individuals to become parents of the next generation. An individual is a parent if it has been selected to undergo variation in order to create offspring. Together with the survivor selection mechanism, parent selection is responsible for pushing quality improvements. In EA, parent selection is typically probabilistic. Thus high quality individuals get a higher chance to become parents than those with low quality. But never the less low quality individuals are often given a small but positive chance, otherwise the whole search could become to greedy and get struck in the local optimum [20].

7.4.5 Variation Operators

The role of variation operators is to create new individuals from old ones. In the corresponding phenotype space this amounts to generating new candidate solutions. From the generate and test search perspective, variation operators perform the generate step. Variation operators are mainly divided into 2 types viz Mutation and Recombination [20].

7.4.5.1 Mutation

Mutation is a genetic operator used to maintain genetic diversity from one generation of a population of genetic algorithm chromosomes to the next. It is analogous to biological mutation. Mutation alters one or more gene values in a chromosome from its initial state. In mutation, the solution may change entirely from the previous solution. Hence GA can come to better solution by using mutation. Mutation occurs during evolution according to a user-definable mutation probability. This probability should be set low. If it is set too high, the search will turn into a primitive random search.

The classic example of a mutation operator involves a probability that an arbitrary bit in a genetic sequence will be changed from its original state. A common method of implementing the mutation operator involves generating a random variable for each bit in a sequence. This random variable tells whether or not a particular bit will be modified. This mutation procedure, based on the biological point mutation, is called single point mutation. Other types are inversion and floating point mutation. When the gene encoding is restrictive as in permutation problems, mutations are swaps, inversions and scrambles.

The purpose of mutation in GAs is preserving and introducing diversity. Mutation should allow the algorithm to avoid local minima by preventing the population of chromosomes from becoming too similar to each other, thus slowing or even stopping evolution. This reasoning also explains the fact that most GA systems avoid only taking the fittest of the population in generating the next but rather a random (or semi-random) selection with a weighting toward those that are fitter [20].

7.4.5.2 Recombination

In EAs, crossover is a genetic operator used to vary the programming of a chromosome or chromosomes from one generation to the next. It is analogous to reproduction and biological crossover, upon which evolutionary algorithms are based. Cross over is a process of taking more than one parent solutions and producing a child solution from them. There are methods for selection of the chromosomes [20].

7.4.6 Survivor Selection Mechanism (Replacement)

The role of Survivor Selection or environmental selection is to distinguish among individuals based on their quality. In that it is similar to parent selection, but it is used in a different stage of the evolutionary cycle. The survivor selection mechanism is called after having created the offspring of the selected parents. Survivor selection is also called replacement or replacement strategy. In EC the population size is always constant, thus the choice has to be made on which individuals will be allowed in the next generation. This decision is usually based on their fitness values, favoring those with higher quality. As opposed to parent selection, survivor selection is often deterministic [20].

7.4.7 Initialization

Initialization is kept simple in most EA applications. The first population is seeded by randomly generated individuals. In principle problem specific heuristics can be used in the step aiming at the initial population with higher fitness. Whether this is worth the extra computational effort or not is very much depending on the application at hand. There are however some general observations concerning this issue based on the so called any time behavior of the EA's [20].

7.4.8   Termination Condition

As for a suitable termination condition we can distinguish 2 cases. If the problem has a known optimal fitness level, probable coming from a known optimum of the given objective function, and then reaching this level should be used as stopping condition. Commonly used options for this purpose are the following [20].

- The maximally allowed CPU time elapses.
- The total number of fitness evolutions reaches a given limit.
- For a given period of time, the fitness improvement remains under a threshold value.
- The population diversity drops under a given threshold.

7.5   Why EA Is Beneficial in Big Data Analysis

Evolutionary algorithms have several features that make them attractive for the Big data handling/data mining process. They are a domain independent technique, which makes them ideal for applications where domain knowledge is difficult to provide. They have the ability to explore large search spaces finding consistently good solutions. In addition, they are relatively insensitive to noise, and can manage attribute interaction better than the conventional data mining techniques.

EAs are different from classical search and optimization procedures in a variety of ways.

Most classical point-by-point algorithms use a deterministic procedure for approaching the optimum solution. Such algorithms start from random guess solution. Thereafter, based on a pre-specified transition rule, the algorithm suggests a search direction, which is often arrived at by considering local information. A uni-directional search is then performed along the search direction to find the best solution. This best solution becomes the new solution and the above procedure is continued for a number of times. Algorithms vary mostly in the way the search directions are defined at each intermediate solution [17].

Every classical optimization algorithm is designed to solve a specific type of problem. For example, the geometric programming method which is designed to solve only polynomial type objective function and constraints. So this geometric programming method is efficient for solving only this type of problems where its not efficient for other type of functions. Thus one algorithm may be best suited for one problem but may not be applicable for a different type of problem. This requires users to know a number of optimization algorithms in order to solve different optimization problems [17].

When the dataset scales up, those algorithms will suffer for the excessive computation cost and the system will retard due to its heavy scan of the large database. When the algorithms are used in Big data applications, the bottleneck becomes more prominent.

Most of the above limitations can be easily avoided by using the Evolutionary algorithms which have the following basic features.

7.5.1 Population Versus Single Best Solution

One of the most striking advantages of EA compared to the classical approach is EAs use a population of solutions in each iteration instead of a single solution. If an optimization problem has a single optimum, all EA population members can be expected to converge to that optimum solution. However, if an optimization problem has multiple optimal solutions, an EA can be used to capture multiple optimal solutions in its final population. This ability of an EA to find multiple optimal solutions in one single simulation run makes EAs unique in solving multi-objective optimization problems. As the 1st step for ideal strategy for multi objective optimization requires multiple trade off solutions, an EA's population approach is the best one which can be used to find a number of solutions in a single simulations run.

The use of a population of solutions helps the evolutionary algorithm avoid becoming "trapped" at a local optimum, when an even better optimum may be found outside the vicinity of the current solution. Considering the volume of data, it's a big advantage, otherwise it can be trapped at a local optimum and it becomes quite difficult to find the global optimum.

7.5.2 Random Versus Deterministic Operation

It relies in part on random sampling. This makes it a nondeterministic method, which may yield somewhat different solutions on different runs—even if we haven't changed our model. In contrast, the linear, nonlinear and integer Solvers also included in the Premium Solver are deterministic methods—they always yield the same solution if we start with the same values in the decision variable cells. Considering the Big data platform, its quite advantageous to explore the search space.

7.5.3 Creating New Solutions Through Mutation

Inspired by the role of mutation of an organism's DNA in natural evolution—an evolutionary algorithm periodically makes random changes or mutations in one or more members of the current population, yielding a new candidate solution (which may be better or worse than existing population members).

If we consider the Big data environment, it provides us the opportunity to explore the search area more successfully by creating new solutions by Mutation.

There are many possible ways to perform a "mutation," and the Evolutionary Solver actually employs three different mutation strategies. The result of a mutation may be an infeasible solution, and the Evolutionary Solver attempts to "repair" such a solution to make it feasible; this is sometimes, but not always, successful.

7.5.4 Combining Solutions Through Crossover

Crossover is inspired by the role of sexual reproduction in the evolution of living things. Here an evolutionary algorithm attempts to combine elements of existing solutions in order to create a new solution, with some of the features of each "parent." The elements (e.g. decision variable values) of existing solutions are combined in a "crossover" operation, inspired by the crossover of DNA strands that occurs in reproduction of biological organisms.

As with mutation, there are many possible ways to perform a crossover operation—some much better than others—and the Evolutionary Solver actually employs multiple variations of two different crossover strategies.

The crossover operation also provides us the opportunity to explore the search area in a more fruitful way by creating new solutions by this operation. Selecting solutions via Survival of the fittest: Inspired by the role of natural selection in evolution—an evolutionary algorithm performs a selection process in which the "most fit" members of the population survive, and the "least fit" members are eliminated. In a constrained optimization problem, the notion of "fitness" depends partly on whether a solution is feasible (i.e. whether it satisfies all of the constraints), and partly on its objective function value. The selection process is the step that guides the evolutionary algorithm towards ever-better solutions.

7.5.5 Independence from the Internal Structure of the Problem

EAs have a great impact on the problems where a differentiable function is difficult to build either due to paucity of sufficient information or due to highly complex nature of the problem. It is happened since the EA does not require the prior knowledge of the problem or its application. We can solve the optimization problem with the help of EA(specifically by GA) provided the problem can be described properly in terms of the chromosome encoding.

7.5.6 Customization of the Algorithm Depending on the Problem Structure

The algorithm can be modified according to the problem structure like it can be scaled well to higher dimensional problems by changing a small parameter easily. Also variation of results can be achieved by changing the selection operators or crossover methods. In the Big data environment, various types of data are available like various unstructured data, images where variation of algorithm is a big advantage.

7.5.7 Simplicity

Another important feature of the EA is its simplicity which has made it so popular. Also it is widely applicable since it is not restricted only to the continuous search spaces.

7.5.8 Drawbacks

A drawback of any evolutionary algorithm is that a solution is "better" only in comparison to other, presently known solutions; such an algorithm actually has no concept of an "optimal solution," or any way to test whether a solution is optimal. (For this reason, evolutionary algorithms are best employed on problems where it is difficult or impossible to test for optimality.) This also means that an evolutionary algorithm never knows for certain when to stop, aside from the length of time, or the number of iterations or candidate solutions, that you wish to allow it to explore.

7.6 When Are Evolutionary Algorithms Useful

Evolutionary algorithms are typically used to provide good approximate solutions to problems that cannot be solved easily using other techniques. Many optimisation problems fall into this category. It may be too computationally-intensive to find an exact solution but sometimes a near-optimal solution is sufficient. In these situations evolutionary techniques can be effective. Due to their random nature, evolutionary algorithms are never guaranteed to find an optimal solution for any problem, but they will often find a good solution if one exists.

One example of this kind of optimisation problem is the challenge of timetabling. Schools and universities must arrange room and staff allocations to suit the needs of their curriculum. There are several constraints that must be satisfied. A member of staff can only be in one place at a time, they can only teach classes that are in their area of expertise, rooms cannot host lessons if they are already occupied, and classes must not clash with other classes taken by the same students. This is a combinatorial problem and known to be NP-Hard. It is not feasible to exhaustively search for the optimal timetable due to the huge amount of computation involved. Instead, heuristics must be used. Genetic algorithms have proven to be a successful way of generating satisfactory solutions to many scheduling problems.

Evolutionary algorithms can also be used to tackle problems that humans don't really know how to solve. An EA, free of any human preconceptions or biases, can generate surprising solutions that are comparable to, or better than, the best human-generated efforts. It is merely necessary that we can recognize a good solution if it were presented to us, even if we don't know how to create a good solution. In other words, we need to be able to formulate an effective fitness function.

Engineers working for NASA know a lot about physics. They know exactly which characteristics make for a good communications antenna. But the process of designing an antenna so that it has the necessary properties is hard. Even though the engineers know what is required from the final antenna, they may not know how to design the antenna so that it satisfies those requirements.

NASA's Evolvable Systems Group has used evolutionary algorithms to successfully evolve antennas for use on satellites. These evolved antennas have irregular shapes with no obvious symmetry (one of these antennas is pictured below). It is unlikely that a human expert would have arrived at such an unconventional design. Despite this, when tested these antennas proved to be extremely well adapted to their purpose.

7.6.1 Pre-requisites

There are two requirements that must be met before an evolutionary algorithm can be used for a particular problem. Firstly, we need a way to encode candidate solutions to the problem. The simplest encoding, and that used by many genetic algorithms, is a bit string. Each candidate is simply a sequence of zeros and ones. This encoding makes cross-over and mutation very straightforward, but that does not mean that you cannot use more complicated representations. In fact, we will see several instances of more advanced candidate representations in later chapters. As long as we can devise a scheme for evolving the candidates, there really is no restriction on the types that we can use. Genetic programming (GP) is a good example of this. GP evolves computer programs represented as syntax trees.

The second requirement for applying evolutionary algorithms is that there must be a way of evaluating partial solutions to the problem—the fitness function. It is not sufficient to evaluate solutions as right or wrong, the fitness score needs to indicate how right or, if your glass is half empty, how wrong a candidate solution is. So a function that returns either 0 or 1 is useless. A function that returns a score on a scale of 1–100 is better. We need shades of grey, not just black and white, since this is how the algorithm guides the random evolution to find increasingly better solutions.

7.7 Common Types of EA

Different types of evolutionary algorithms have been developed over the years such as genetic algorithms, genetic programming, evolution strategies, evolutionary programming, evolution strategies, differential evolution, cultural evolution algorithms and co-evolutionary algorithms. Some of these types that are used in data mining are genetic algorithms, genetic programming and co-evolutionary algorithms. Genetic algorithms are used for data preprocessing and for post processing the discovered knowledge, while genetic programming is used for rule discovery and data preprocessing. Here we will discuss the features of Genetic Algorithm in the light of EA.

7.7.1 Genetic Algorithms

Genetic algorithm is a probabilistic optimization method based on the natural evolution model, proposed by Holland in 1975. Charles Darwin's famous strategy Survival of the fittest is the key idea which is followed in the GA to find the optimal solution. Genetic algorithm was developed to simulate some of the processes observed in natural evolution, a process that operates on chromosomes (organic devices for encoding the structure of living being). The genetic algorithm differs from other search methods in that it searches among a population of points, and works with a coding of parameter set, rather than the parameter values themselves. It also uses objective function information without any gradient information. The transition scheme of the genetic algorithm is probabilistic, whereas traditional methods use gradient information. Because of these features of genetic algorithm, they are used as general purpose optimization algorithm. They also provide means to search irregular space and hence are applied to a variety of function optimization, parameter estimation and machine learning applications.

7.7.1.1 Basic Version: Simple Genetic Algorithm

The Simple Genetic Algorithm is the basic version of Genetic algorithm which primarily comprises of four basic steps: (1) Initial population generation, (2) Choosing the best parent chromosome with the help of different selection operators by applying the fitness function, (3) Crossover operation to generate new offspring from the parent chromosome and (4) Mutation operation to maintain diversity in the population [18].

In the first step of the GA, a pool of individuals called population is randomly created within the specified search range.

The next step is to choose the selection method and based on that method, two fittest individuals are randomly selected from the population depending on their fitness values. The main principle of selection strategy is the better is an individual; the higher is its chance of being parent.

The next two steps comprise of Crossover and Mutation operations which are generally used to explore the new and unknown areas in the search space where the selection process reduces the search space by eliminating the poor individuals. This process continues until the best solution is achieved [2, 52].

The Simple Genetic Algorithm (SGA) is represented algorithmically as follows:

1. *[Start] Initially generate randomly a population of n individuals/chromosomes (probable solutions for the problem).*
2. *[Fitness] Evaluate the fitness score f(x) of each chromosome x in the population.*
3. *[New population] Create a new population by repeating the following steps.*

   a. *[Elitism] Find the fittest chromosome from the population based on the fitness function value.*
   b. *[Selection] Select randomly two parent chromosomes from a population based on a selection method.*
   c. *[Crossover] Apply crossover between the parents selected in previous step. As a result two new offspring will be generated.*
   d. *[Mutation] Apply Mutation on the offsprings generated in the previous step.*
   e. *[Accepting] Place new offspring in the new population.*

4. *[Replace] Use newly generated population for another execution of the algorithm.*
5. *[Test] If the stopping condition is satisfied then go To Step 6 otherwise go To Step 2.*
6. *[Stop] Terminate the algorithm.*

We have used Elitism as a part of new population creation process in the SGA used for this experiment. The best chromosome of the current population will move to the next generation. The rest of the chromosomes of that new population will be created based on the selection process used in the algorithm.

7.7.1.2 Selection Techniques

In the GA, exploration is the process of visiting entirely new regions of a search space. This is controlled by Crossover and Mutation operators by creating diversity in the population. On the other hand, exploitation which is the reduction of diversity by focusing on the better individuals is taken care by the selection process. As a result, a good search technique must find a good balance between these two important factors: exploration and exploitation to find a global optimum. A trade off ensures that good solutions go to the next generation more frequently than poor solutions. However there is a chance that poor solutions may pass to the next generation.

Various selection strategies are used in the Genetic Algorithm to find out the good solution/best chromosomes. These are Tournament selection, Roulette wheel selection, Boltzman selection, Rank selection, Steady state selection methods.

These selection strategies significantly affect the performance of the algorithm in different manner because of their diverse and implicit nature. It is observed by the researchers that any one technique is not enough to obtain the best optimal solution for all kind of problems since every technique has its own merits and demerits. We have selected two most common and simple selection methods, Tournament selection and Roulette wheel selection processes comparing for developing hybrid selection and enhanced selection processes.

Basic Selection Techniques

Below are the selection techniques which are commonly used for selection operators in the Genetic algorithm. Every selection technique has its own advantages and disadvantages. The different selection strategy used in the GA process will significantly affect the performance of the algorithm differently. According to the previous researches, the Roulette wheel method consumes the highest iteration time whereas Tournament selection performs better than the Roulette wheel selection for most of the test functions including the higher dimensional functions.

- **Roulette Wheel Selection**
- **Tournament Selection**
- **Boltzmann Selection**
- **Rank Selection**
- **Steady state Selection**

We are discussing here about the most commonly used techniques: Roulette Wheel Selection and Tournament Selection.

A. Roulette Wheel Selection

In Roulette wheel selection, the parents are selected according to their fitness values. The chance of selection of chromosome increases with the betterment of chromosome. A Roulette wheel can be visualized where all chromosomes of the population are placed for selection. Every chromosome will increase its share in the wheel according to the fitness value.

Now if a marble is thrown in the wheel, the chance of selecting the fittest chromosomes is more compared to others. The above method works fine for the maximization problems where fitter individual means the individuals with greater fitness value. So according to the Roulette wheel functionality, these fitter individuals take more place in the wheel than the weaker ones.

But the scenario is little bit different if the function is minimization function instead of maximization function. In that case, the fitter individual means the individuals with lesser fitness value. Hence little bit modification was done in the original algorithm to implement it for minimization function. Instead of sum of fitness of the population, the sum of reciprocal of fitness is used.

The basic advantage of the roulette wheel selection is that it discards none of the individuals in the population and gives a chance to all of them to be selected. Therefore, the diversity is maintained in the population.

This algorithm used here is described below.

1. *[Sum] Calculate sum of all chromosome fitnesses in population which is sum S.*
2. *[Select] Generate a random number in the interval (0, S) which is r.*

3. *[Loop] Go through the population and reciprocal sum of fitnesses which is s. When the sum s is greater then r, stop and return the chromosome which is in the current loop.*

Note: step 1 is performed only once for each population.

B. Tournament Selection

Tournament Selection is one of the most widely used selection strategies in evolutionary algorithms. It involves several tournaments among a few individuals chosen at random from the population. The individual with the best fitness value is the winner of that particular tournament and it is selected for other operations like mutation and crossover. Tournament size is used to adjust the selection pressure. If the tournament size increases, the chance of selecting the weaker individuals increases. Tournament selection also gives a chance to most of the individuals to be selected and thus it preserves diversity, although keeping diversity may degrade the convergence speed sometimes [52].

The main advantage of this method is that it allows the selection pressure to be easily adjusted.

The algorithm can be described below.

1. *[Input] Population and Tournament size will be passed as Input values.*
2. *[Select] Generate a random number p in the interval (1, N) where N is the population size. The individual with position p will be selected from the given population for the Tournament.*
3. *[Loop] Go to Step 2 until all the individuals of the Tournament are selected.*
4. *[Select Fittest Individual] Find the best individual of the Tournament based on the fitness value.*

Hybrid Selection Techniques

There are some advantages and disadvantages of all the selection methods. So if we can create a hybrid selection procedure where Roulette wheel and Tournament methods are blended, it can improves the merits of the above two algorithms and reduces their demerit. Primarily two individuals are required to be selected for crossover and mutation operation at a single point of time. Hybrid selection selects one individual by Roulette wheel selection and the other by Tournament selection. The Enhanced Roulette wheel selection technique is basically the modification of traditional Roulette wheel Method. Modification is implemented depending on the value of average fitness deviation of the population. The fitness deviation (deviation from the optimal value) of the individual selected by modified roulette wheel selection should not be greater than the average fitness deviation of the population.

Hybrid Selection

The genetic algorithm is described below after applying the Hybrid selection method.

1. *[Start] Generate random by a population of n chromosomes (probable solutions for the problem).*
2. *[Fitness] Evaluate the fitness f(x) of each chromosome x in the population.*

3. *[New Population] Create a new population by repeating following steps until the new population is generated.*

   a. *[Elitism] Find the fittest chromosome from the population based on the fitness value and it will be placed to the next generation population.*
   b. *[Hybrid Selection] Select one parent chromosome from the population based on Tournament selection and the other by the Roulette-wheel selection.*
   c. *[Crossover] Apply crossover between the parents selected through previous step. As a result a new offspring will be generated.*
   d. *[Mutation] Apply Mutation on the offsprings generated in the previous step.*
   e. *[Accepting] Place new offspring in the new population.*

4. *[Replace] Find the best individual of the Tournament based on the fitness value.*
5. *[Test] Population and Tournament size will be passed as Input values.*
6. *[Loop] Generate a random number p in the interval (1, N) where N is the population size. The individual with position p will be selected from the given population for the Tournament.*

Enhanced Roulette wheel Selection
The modified Roulette wheel selection process is described below.

1. *[Sum] Calculate sum of all chromosome fitnesses in the population which is S*
2. *[Total Fitness Deviation] Calculate the sum of fitness deviation for all the individuals of the population.*
3. *[Average Fitness Deviation] Calculate the average fitness deviation, dividing the Total Fitness Deviation by the number of individuals of the population.*
4. *[Select] Generate a random number in the interval (0, S) which is r.*
5. *[Loop] Go through the population and the reciprocal sum of fitnesses which is s. When the sum s is greater than r, stop and return the chromosome which is in the current loop.*
6. *[Check] Check whether the fitness deviation of the selected individual is lesser than the average fitness deviation or not.*
7. *[Accept] If the fitness deviation of the selected individual is lesser than the average fitness deviation, accept the solution, otherwise go to step 4.*

Note: step 1, 2 and 3 are performed only once for each population.

7.7.1.3   Application of Genetic Algorithm in the Light of Big Data

At this time, Big data applications are becoming the main focus of attention because of the huge increase in data generation and storage that has occurred in recent years. This situation becomes a challenge when huge amounts of data are processed to extract knowledge because data mining techniques are not adapted to the new requirements of space and time.

With the huge amount of data being generated in the world every day, at a rate far higher than by which it can be analysed by human comprehension alone, data mining becomes an extremely important task for extracting as much useful information from this data as possible. The standard data mining techniques are satisfactory to a certain extent but they are constrained by certain limitations, and it is for these cases that evolutionary approaches are both more capable and more efficient.

As computers invaded almost all conceivable fields of human knowledge and occupation, their advantages were advocated all over, but what was observed soon enough was that with the increasing amounts of data that could be generated, stored and analysed there was a need to define some way to sift through it and grab the important stuff out. During the earlier days a human or a group of humans would sit down to analyse the data by going through it manually and using statistical techniques, but the curve of data generation was far steeper than what could realistically be processed by hand. This led to the emergence of the field of data mining, which was essentially to define and formalize standard techniques to extract data from large data warehouses. As data mining evolved it was observed that the data at hand was almost always never perfect or suitable to be fed to data mining engines and needed several steps of pre-processing before it could be put through mining. Generally these inconsistencies would be in data format, level of noise or incorrect data, unnecessary data, redundant data etc. These steps would clean, integrate, discretize and select the most relevant attributes before performing any mining.

A whole new area called Intelligent data analysis has emerged which utilizes efficient techniques for mining data from large sets keeping in mind that the knowledge obtained is useful at the same time also remembering that time for mining is constrained and the user requires data as soon as possible. Some of the methods used to mine data include support vector machines, decision trees, nearest neighbour analysis, Bayesian classification, and latent semantic analysis.

With the problems associated with conventional data mining techniques, clever new ways to overcome these were needed, and the application of AI techniques to the field resulted in a very powerful hybrid of techniques. Evolutionary optimization techniques provided with a useful and novel solution to these issues, and once data mining was enhanced with using EC many of the previously mentioned problems were no longer big issues. Here we will discuss some of the areas where these are implemented and give a few ideas of where these techniques may be implemented in the future. Evolutionary algorithms have several features that make them attractive for the data mining process [23, 51]. They are a domain independent technique, which makes them ideal for applications where domain knowledge is difficult to provide.

They have the ability to explore large search spaces finding consistently good solutions. In addition, they are relatively insensitive to noise, and can manage attribute interaction better than the conventional data mining techniques. Therefore, several works have been done, in recent years, to develop new techniques for data mining using evolutionary algorithms. These attempts used evolutionary algorithms for different tasks of data mining such as feature extraction, feature selection, classification, and clustering [10]. The main role of evolutionary algorithms in most of these approaches is optimization. They are used to improve the robustness and accuracy of some of the traditional data mining techniques.

Below are some real life applications where Genetic algorithm (a kind of Evolutionary algorithm) can be used as a feasible solution compared to other classical methods as all these applications deal with large amount of complex data set.

Genetic Algorithm—Effective tool in Data Mining and Pattern Recognition:
Data mining algorithms require a technique that partitions the domain values of an attribute in a limited set of ranges, simply because considering all possible ranges of domain values is infeasible [52].

There are two different methods to applying GA in pattern recognition:

1. *Use GA as a classifier directly in computation.*
2. *Use a GA to optimize the results i.e. as an optimizer to arrange the parameters in other classifiers.*

Most applications of GAs in pattern recognition optimize some parameters in the classification process [4].

GAs has been applied to find an optimal set of feature weights that improve classification accuracy. First, a traditional feature extraction method such as Principal Component Analysis (PCA) is applied, and then a classifier such as k-NN (Nearest Neighbour Algorithm) is used to calculate the fitness function for GA [41, 46]. Combination of classifiers is another area that GAs have been used to optimize. GA is also used in selecting the prototypes in the case-based classification

According to us second method of genetic algorithm to optimize the result from the dataset is more effective to compute the accurate values of observations of data by applying data mining techniques.

Genetic Algorithm—wide scope in business:
There are large amount of data that has to be filtered to process the results for optimizing the business profits by using various data mining techniques. There are many domains in business to which they can be applied:

I. Optimization
Give a business problem with certain variables and a well-defined definition of profit, a genetic algorithm can be used to automatically determine the optimal value for the variables that optimize the profit [22].

II. Prediction
Genetic algorithms have been used as meta level operators that are used to help optimize other data mining algorithms. For instance, they have been used to find the optimal association rules in market-analysis.

III. Simulation
Sometimes a specific business problem is not well defined in terms what the profit is or whether one solution is better than the other. The business person instead just has large number of entities that they would like to simulate via simple interaction rules overtime.

Genetic Algorithm—stock exchange data mining:
Stock market and other finance fields, Genetic Algorithm has been applied in many problems [11]. There have been a number of attempts to use GA for acquiring technical trading rules.

One application is how to find the best combination values of each parameter. We know that in a trading rule there are many parameters, when we try to find the most profit, we must test the parameter combination one by one, which is called greedy algorithm which costs a lot of running time and memory.

Genetic Algorithm in light of information theory:
The field of Information Theory refers Big data as datasets whose rate of increase is exponentially high and in small span of time; it becomes very painful to analyse them using typical data mining tools. Such data sets results from daily capture of stock exchange, any credit card user's timely usage trends, insurance cross line capture, health care services etc. In real time these data sets go on increasing and with passage of time create complex scenarios. Thus the typical data mining tools needs to be empowered by computationally efficient and adaptive technique to increase degree of efficiency by using adaptive techniques.

Using GA over data mining creates great robust, computationally efficient and adaptive systems. In past there have been several researches on data mining using statistical techniques. The statistics that have heavily contributed are the ANOVA, ANCOVA, Poisson's Distribution, and Random Indicator Variables etc. The biggest drawback of any statistical tactics lies in its tuning. With exponential explosion of data, this tuning goes on taking more time and inversely affects the through put. Also due to their static nature, often complex hidden patterns are left out. The idea here is to use genes to mine out data with great efficiency.

Genetic Algorithm—Automotive Design:
Using Genetic Algorithms [GAs] to both design composite materials and aerodynamic shapes for race cars and regular means of transportation (including aviation) can return combinations of best materials and best engineering to provide faster, lighter, more fuel efficient and safer vehicles for all the things we use vehicles for. Rather than spending years in laboratories working with polymers, wind tunnels and balsa wood shapes, the processes can be done much quicker and more efficiently by computer modelling using GA searches to return a range of options human designers can then put together however they please.

Design processes of modern race car are often developed in short time; during this period a large number of parameters has to be tuned to reach best results. Much kind of vehicle dynamics simulation models have been developed by car manufacturers and private suppliers to investigate car behaviour on racing tracks. Such models have a high degree of complexity but they can be employed, in a simplified mode, during design processes of racing cars for which tracks technical data (for such circuits where they will race) are well known. During the first stages of the design activity very complex numerical models (such as multibody simulations) are not necessary and it is possible to use simplified methods to locate optimal solutions in a fast way. In present work a numerical model, able to reproduce car behaviour on a defined circuit or simply analyse meaningful test cases (breaking, steering,), is used to appraise performances of race car with different technical configuration. Besides is proposed a method to reduce designing field of investigation through genetic algorithms (GA's) using vehicle numerical model to individuate such solution that gives best performances on each circuit. Numerical algorithm is tested to choose best from a large number of technical configurations on different tracks. Several types of genetic algorithms are used: results show that GA's can be a useful tool to fasten design processes narrowing the field of investigation. In some cases the optimization method developed gives analogous results as what would be carried out by an experienced engineer.

Genetic Algorithm—Engineering Design:
Getting the most out of a range of materials to optimize the structural and operational design of buildings, factories, machines, etc. is a rapidly expanding application of GAs. These are being created for such uses as optimizing the design of heat exchangers, robot gripping arms, satellite booms, building trusses, flywheels, turbines, and just about any other computer-assisted engineering design application. There is work to combine GAs optimizing particular aspects of engineering problems to work together, and some of these can not only solve design problems, but also project them forward to analyze weaknesses and possible point failures in the future so these can be avoided.

Genetic Algorithm—Field of Robotics:
Robotics involves human designers and engineers trying out all sorts of things in order to create useful machines that can do work for humans. Each robot's design is dependent on the job or jobs it is intended to do, so there are many different designs out there. GAs can be programmed to search for a range of optimal designs and components for each specific use, or to return results for entirely new types of robots that can perform multiple tasks and have more general application.

Genetic Algorithm—Evolvable Hardware:
Evolvable hardware (EH) is a new field about the use of evolutionary algorithms (EA) to create specialized electronics without manual engineering. It brings together reconfigurable hardware, artificial intelligence, fault tolerance and autonomous systems. Evolvable hardware refers to hardware that can change its architecture and behaviour dynamically and autonomously by interacting with its environment.

Evolvable hardware applications are electronic circuits created by GA computer models that use stochastic (statistically random) operators to evolve new configurations from old ones. As the algorithm does its thing in the running model, eventually a circuit configuration will come along that does what the designer wants. Think of reconfigurable circuits in something like a space robot. It could use a built-in GA library and simulator to re-design itself after something like radiation exposure that messes up its normal configuration, or encounters a novel situation in which it needs a function it doesn't already have. Such GAs would enable self-adaptation and self-repair.

In its most fundamental form an evolutionary algorithm manipulates a population of individuals where each individual describes how to construct a candidate circuit.

Each circuit is assigned a fitness, which indicates how well a candidate circuit satisfies the design specification. The evolutionary algorithm uses stochastic operators to evolve new circuit configurations from existing ones. Done properly, over time the evolutionary algorithm will evolve a circuit configuration that exhibits desirable behaviour.

Each candidate circuit can either be simulated or physically implemented in a reconfigurable device. Typical reconfigurable devices are field-programmable gate arrays (for digital designs) or field-programmable analog arrays (for analog designs). At a lower level of abstraction are the field-programmable transistor arrays that can implement either digital or analog designs.

The concept was pioneered by Adrian Thompson at the University of Sussex, England, who in 1996 evolved a tone discriminator using fewer than 40 programmable logic gates and no clock signal in a FPGA. This is a remarkably small design for such a device and relied on exploiting peculiarities of the hardware that engineers normally avoid. For example, one group of gates has no logical connection to the rest of the circuit, yet is crucial to its function.

In many cases, conventional design methods (formulas, etc.) can be used to design a circuit. But in other cases, the design specification doesn't provide sufficient information to permit using conventional design methods. For example, the specification may only state desired behaviour of the target hardware.

In other cases, an existing circuit must adapti.e., modify its configurationto compensate for faults or perhaps a changing operational environment. For instance, deepspace probes may encounter sudden high radiation environments, which alter a circuit's performance; the circuit must self-adapt to restore as much of the original behaviour as possible.

The fitness of an evolved circuit is a measure of how well the circuit matches the design specification. Fitness in evolvable hardware problems is determined via two methods:

- extrinsic evolution: all circuits are simulated to see how they perform,
- intrinsic evolution: physical tests are run on actual hardware.

In extrinsic evolution, only the final best solution in the final population of the evolutionary algorithm is physically implemented, whereas with intrinsic evolution every individual in every generation of the EA's population is physically realized and tested.

Genetic Algorithm—Optimized Telecommunications Routing:
A telecommunications network is a collection of terminal nodes, links and any intermediate nodes which are connected so as to enable telecommunication between the terminals.

The transmission links connect the nodes together. The nodes use circuit switching, message switching or packet switching to pass the signal through the correct links and nodes to reach the correct destination terminal.

Each terminal in the network usually has a unique address so messages or connections can be routed to the correct recipients. The collection of addresses in the network is called the address space.

Examples of telecommunications networks are:

- computer networks
- the Internet
- the telephone network
- the global Telex network
- the aeronautical ACARS network

Sometimes we get frustrated by slow LAN performance, inconsistent internet access, a FAX machine that only sends faxes sometimes, our land line's number of 'ghost' phone calls every month. GAs are being developed that will allow for dynamic and anticipatory routing of circuits for telecommunications networks. These could take notice of our system's instability and anticipate our re-routing needs. Using more than one GA circuit-search at a time, soon our interpersonal communications problems may really be all in our head rather than in our telecommunications system. Other GAs are being developed to optimize placement and routing of cell towers for best coverage and ease of switching.

Genetic Algorithm—Joke and Pun Generation:
Among the linguistic applications of GAs—including a JAPE (automated pun generator) inspired STANDUP program to design communications strategies for people working with children who suffer communications disabilities—are GAs that search for jokes and puns. These come under the heading of "artificial creativity" and AI, but could prove very useful to class clowns and wannabe punsters whose public reputations depend upon being funnier than they actually are.

Computational creativity (also known as artificial creativity, mechanical creativity or creative computation) is a multidisciplinary endeavor that is located at the intersection of the fields of artificial intelligence, cognitive psychology, philosophy, and the arts.

The goal of computational creativity is to model, simulate or replicate creativity using a computer, to achieve one of several ends:

- To construct a program or computer capable of human-level creativity.
- To better understand human creativity and to formulate an algorithmic perspective on creative behavior in humans.
- To design programs that can enhance human creativity without necessarily being creative themselves.

The field of computational creativity concerns itself with theoretical and practical issues in the study of creativity. Theoretical work on the nature and proper definition of creativity is performed in parallel with practical work on the implementation of systems that exhibit creativity, with one strand of work informing the other.

Genetic Algorithm—Biomimetic Invention:
Biomimicry or biomimetics is the development of technologies inspired by designs in nature. Since GAs are inspired by the mechanisms of biological evolution, it makes sense that they could be used in the process of invention as well. GAs rely primarily on something called implicit parallelism (like to like), using mutation and

selection in secondary roles toward a design solution. GA programmers are working on applications that not only analyze the natural designs themselves for a return on how they work, but can also combine natural designs to create something entirely new that can have exciting applications.

Genetic Algorithm—Trip, Traffic and Shipment Routing:
New applications of a GA known as the "Traveling Salesman Problem" or TSP can be used to plan the most efficient routes and scheduling for travel planners, traffic routers and even shipping companies. The shortest routes for traveling. The timing to avoid traffic tie-ups and rush hours. Most efficient use of transport for shipping, even to including pickup loads and deliveries along the way. The program can be modeling all this in the background while the human agents do other things, improving productivity as well! Chances are increasing steadily that when you get that trip plan packet from the travel agency, a GA contributed more to it than the agent did.

Genetic Algorithm—Computer Gaming:
Those who spend some of their time playing computer Sims games (creating their own civilizations and evolving them) will often find themselves playing against sophisticated artificial intelligence GAs instead of against other human players online. These GAs have been programmed to incorporate the most successful strategies from previous games—the programs 'learn'—and usually incorporate data derived from game theory in their design. Game theory is useful in most all GA applications for seeking solutions to whatever problems they are applied to, even if the application really is a game.

Genetic Algorithm—Encryption and Code Breaking:
On the security front, GAs can be used both to create encryption for sensitive data as well as to break those codes. Encrypting data, protecting copyrights and breaking competitors' codes have been important in the computer world ever since there have been computers, so the competition is intense. Every time someone adds more complexity to their encryption algorithms, someone else comes up with a GA that can break the code. It is hoped that one day soon we will have quantum computers that will be able to generate completely indecipherable codes. Of course, by then the 'other guys' will have quantum computers too, so it's a sure bet the spy vs. spy games will go on indefinitely.

Genetic Algorithm—Computer-Aided Molecular Design:
The de novo design of new chemical molecules is a burgeoning field of applied chemistry in both industry and medicine. GAs are used to aid in the understanding of protein folding, analyzing the effects of substitutions on those protein functions, and to predict the binding affinities of various designed proteins developed by the pharmaceutical industry for treatment of particular diseases. The same sort of GA optimization and analysis is used for designing industrial chemicals for particular uses, and in both cases GAs can also be useful for predicting possible adverse consequences. This application has and will continue to have great impact on the costs associated with development of new chemicals and drugs.

Genetic Algorithm—Gene Expression Profiling:
The development of microarray technology for taking 'snapshots' of the genes being expressed in a cell or group of cells has been a boon to medical research. GAs has been and is being developed to make analysis of gene expression profiles much quicker and easier. This helps to classify what genes play a part in various diseases, and further can help to identify genetic causes for the development of diseases. Being able to do this work quickly and efficiently will allow researchers to focus on individual patients' unique genetic and gene expression profiles, enabling the hoped-for "personalized medicine" we've been hearing about for several years.

Genetic Algorithm—Optimizing Chemical Kinetic Analysis:
In the not-so rarified realm of fuels and engines for combustion technologies, GAs are proving very useful toward optimizing designs in transportation, aerospace propulsion and electrical generation. By being able to predict ahead of time the chemical kinetics of fuels and the efficiency of engines, more optimal mixtures and designs can be made available quicker to industry and the public. Some computer modeling applications in this area also simulate the effectiveness of lubricants and can pinpoint optimized operational vectors, and may lead to greatly increased efficiency all around well before traditional fuels run out.

Genetic Algorithm—Finance and Investment Strategies:
In the current unprecedented world economic meltdown one might legitimately wonder if some of those Wall Street gamblers made use of GA-assisted computer modeling of finance and investment strategies to funnel the world's accumulated wealth into what can best be described as dot-dollar black holes. But then again, maybe they were simply all using the same prototype, which hadn't yet been de-bugged. It may be possible that a newer generation of GA-assisted financial forecasting would have avoided the black holes and returned something other than bad debts the taxpayers get to repay.

Genetic Algorithm—Marketing and Merchandising:
We could think the word 'merchandising' just the way Mel Brooks said it in the "Space Balls" the movie. Space Balls the toilet paper. Space Balls the lunchbox. Space Balls the flame thrower (the kids love this one)... And laugh because it's close enough to reality to be funny. So it shouldn't surprise anyone that GAs are indeed being put to work to help merchandisers to produce products and marketing consultants design advertising and direct solicitation campaigns to sell stuff. Maybe this application of GAs could someday get us out of the financial black hole and get things moving again.

7.7.2 Particle Swarm Optimization

Data mining has been a popular academic topic in computer science and statistics for decades, swarm intelligence is a relatively new subfield of computational intelligence

(CI) which studies the collective intelligence in a group of simple individuals. In the swarm intelligence, useful information can be obtained from the competition and cooperation of individuals. Generally, there are two kinds of approaches that apply swarm intelligence as data mining techniques [45]. The first category consists of techniques where individuals of a swarm move through a solution space and search for solution(s) for the data mining task. This is a search approach; the swarm intelligence is applied to optimize the data mining technique, e.g., the parameter tuning. In the second category, swarms move data instances that are placed on a low-dimensional feature space in order to come to a suitable clustering or low-dimensional mapping solution of the data. This is a data organizing approach; the swarm intelligence is directly applied to the data samples, e.g., dimensionality reduction of the data. Swarm intelligence, especially particle swarm optimization or ant colony optimization algorithms, is utilized in data mining to solve single objective [1] and multi-objective problems [15]. Based on the two characters of particle swarm, the self-cognitive and social learning, the particle swarm has been utilized in data clustering techniques [45], document clustering, variable weighting in clustering high-dimensional data [39], semi-supervised learning based text categorization, and the Web data mining [50]. In a swarm intelligence algorithm, there are several solutions exist at the same time. The premature convergence may happen due to the solution getting clustered together too fast. However, the solution clustering is not always harmful for optimization. In a brain storm optimization algorithm, the clustering analysis is utilized to reveal the landscapes of problems and to guide the individuals to move toward the better and better areas [54]. Every individual in the brain storm optimization algorithm is not only a solution to the problem to be optimized, but also a data point to reveal the landscapes of the problem [9]. The machine learning and data mining techniques can be combined to produce benefits above and/or beyond what either method could achieve alone [40]. The Big data analytics is required to manage immense amounts of data quickly [16]. The amount of data are attracting more and more attentions, however, the dimension of data and the number of objective of problems also increase the hardness of problems. Three kinds of difficulties should be overwhelmed to solve Big data problems:

7.7.2.1 Large Scale Optimization

The Big data analytics requires a fast mining on the large scale dataset, i.e., the immense amounts of data should be processed in a limited time. The analytic Swarm Intelligence in Big data Analytics problem can be modelled as optimization problems. In general, optimization concerns with finding the best available solution(s) for a given problem within allowable time, and the problem may have several or numerous optimum solutions, of which many are local optimal solutions. Normally, the difficulty of problem will increase with the increasing of the number of variables and objectives. Specially, problems with large number of variables, e.g., more than thousands variables, are termed as large scale problems. Many optimization methods suffer from the curse of dimensionality, which implies that their performance deteriorates quickly as the dimension of the search space increases [5, 14, 29]. There are several reasons that cause this phenomenon. First, the solution space of a problem often increases exponentially with the problem dimension and more efficient search strategies are required to explore all promising regions within a given time budget. The evolutionary computation or swarm intelligence is based on the interaction of a group of solutions. The promising regions or the landscape of problems are very difficult to reveal by small solution samples (compared with the number of all feasible solutions). The empty space phenomenon gives an example of problems getting hard when the dimension increases [40]. The number of possible solutions is increased exponentially when the dimension increasing. The search performance of most algorithms is based on the previous search experience. Considered the limitation of computational resources, the percentages of data points have been retrieved will close to zero when the dimension increased to a large number. The performance of algorithms is affected by the increasing of problems' dimension. Second, the characteristics of a problem may change with the scale. Problems will become more difficult and complex when the dimension increases. Rosen- brock's function, for instance, is unimodal for two dimensional problems but becomes multimodal for higher dimensional problems. Because of such a worsening of the features of an optimization problem resulting from an increase in scale, a previously successful search strategy may no longer be capable of finding an optimal solution. Third, the direction of good solutions is difficult to determine. The swarm intelligence takes an update on each dimension, evaluation on whole dimensions strategy. An algorithm is very difficult to determine which one is better when two solutions both have some good parts and their fitness values are equally bad. The similar scenario also happens in multiobjective optimization. In Pareto domination measurement, nearly all solutions are Pareto non-dominated when the number of objects is larger than 10. The last, the bias is accumulated. In the swarm intelligence, each solution is updated dimension by dimension, and the fitness value is calculated for the whole solution. The solution update depends on the combination of several vectors, i.e., the current value, the difference between current value and previous best value, the differential between current value and neighbour best value, or the difference between two random solutions, etc. In the low dimensional space, the direction of the vector combination has the high probability to point to the global optimum. However, the distance metric, which is utilized in low dimension space, is not S. Cheng et al. effective in high dimensional space. The search direction is far away from the global optimum due to the bias accumulation. Many effective strategies are proposed for high dimensional optimization problems, such as problem decomposition and subcomponents cooperation, parameter adaptation, surrogate-based fitness evaluations [38]. Based on the swarm intelligence, an effective method could find good solutions for large scale problems, both on the time complexity and result accuracy.

7.7.2.2 Handling High Dimensional Data

The curse of dimensionality also happens on the high dimensional data min- ing problems [29, 32, 35]. Many algorithms' performance deteriorates quickly as the dimension of the data space increases. For example, the nearest neighbour approaches are very effective in categorization. However, for high dimensional data, it is very difficult to solve the similarity search problem due to the computational complexity, which was caused by the increase of dimensionality. Many methods are proposed on the high dimension data mining problems. Transforming the high dimensional mining problems into low dimensional space via a projection operation is an effective way. The locality sensitive hashing algorithm is proposed to find nearest neighbours in the high dimensional space [16]. This algorithm is based on hashing functions with strong local-sensitivity in order to retrieve nearest neighbours in a Euclidean space with a complexity sub linear in the amount of data. The data mining problem can be transformed as an optimization problem, because many researchers have been taken o n the large scale optimization problems. Swarm intelligence, especially particle swarm optimization or ant colony optimization algorithms, is utilized in data mining to solve single objective [1] and multiobjective problems [13].

7.7.2.3 Handling Dynamical Data

The Big data, such as the web usage data of Internet, real time traffic information, rapidly changes over time. The analytical algorithms need to process these data swiftly. The dynamic problems, sometimes termed as non-stationary environments, or uncertain environments [37], dynamically change over time. Swarm intelligence has been widely applied to solve stationary and dynamical optimization problems. Swarm intelligence often has to solve optimization problems in the presence of a wide range of uncertainties. Generally, uncertainties in optimized problems can be divided into the following categories.

1. The fitness function or the processed data is noisy.
2. The design variables and/or the environmental parameters may change after opti- mization, and the quality of the obtained optimal solution should be robust against environmental changes or deviations from the optimal point Swarm Intelligence in Big data Analytics.
3. The fitness function is approximated [38], such as surrogate-based fitness evalu- ations, which means that the fitness function suffers from approximation errors.
4. The optimum in the problem space may change over time. The algorithm should be able to track the optimum continuously.
5. The target of optimization may change over time. The demand of optimization may adjust to the dynamical environment, for example, there should be a balance between the computing efficiency and the computational cost for different com- puting loads. In all these cases, additional measures must be taken so that swarm intelligence algorithms are still able to solve satisfactorily dynamic problems [37].

7.7.2.4   Multi-objective Optimization

Different sources of data are integrated in the Big data research, and in most of the Big data analytics problems, more than one objective need to be satisfied at the same time. According to the number of objectives, optimization problems can be divided as single objective and multiobjective problems. For the multi-objective problems, the traditional mathematical programming techniques have to perform a series of separate runs to satisfy different objectives. Multiobjective Optimization refers to optimization problems that involve two or more objectives, and a set of solutions is sought instead of one. A general multiobjective optimization problem can be described as a vector function f that maps a tuple of n parameters (decision variables) to a tuple of k objectives. Unlike the single objective optimization, the multiobjective problems have many or infinite solutions [8].

The optimization goal of an MOP consists of three objectives:

1. The distance of the resulting nondominated solutions to the true optimal Pareto front should be minimized;
2. A good (in most cases uniform) distribution of the obtained solutions is desirable;
3. The spread of the obtained nondominated solutions should be maximized, i.e., for each objective a wide range of values should be covered by the nondominated solutions.

In a multiobjective optimization problem, we aim to find the set of optimal trade-off solutions known as the Pareto optimal set. Pareto optimality is defined with respect to the concept of nondominated points in the objective space. Swarm intelligence methods can effectively solve the multiobjective problems. Several new techniques are combined in the swarm intelligence techniques to solve multiobjective problems with more than ten objectives, in which almost every solution is Pareto nondominated in the problems. These techniques include objective decomposition, objective reduction [30], and clustering in the objective space [54].

7.7.2.5   Applications

The Big data is created in many fields in everyday life. With the Big data analytics techniques and swarm intelligence methods, more effective applications or systems can be designed to solve real world problems. The intelligent transportation system and wireless sensor networks are two typical examples of Big data analytics application.

Intelligent Transportation System

The traffic problems are arising in many cities now. The traffic and transportation system is affected by many factors, such as the number of vehicles, weather, accidents, etc., and the traffic information changes in real time. The purpose of intelligent transport is to build more rapid, safe, and more efficient traffic and transportation systems by constructing the intelligent vehicles and road environment [49]. There are more than one objectives which need to be satisfied at the same time in intelligent transport systems, for example, rapid transportation, environmental pollution, transportation scheduling; and many of these objectives are conflicted with each other.

Wireless Sensor Networks
Based on the wireless sensor networks, the physical world is turning to be a kind of information system [12]. Different sensors are connected to form a net- works; information is transferred in this network by communication techniques. The physical world's information from sensor networks can be collected almost anywhere at any time. The sensor networks and communication techniques have constructed a new paradigm, which is called the internet of things [12]. The wireless sensor networks have been applied to many real-world problems, such as environmental surveillance, transportation monitoring, engineering surveying, and industrial control, just to name a few [42]. Massive data will be generated from the long term and/or large scale wireless sensor network system. The goal of data analysis is to make the fastest possible revelation toward the useful information. Swarm intelligence is an effective way to handle these data, and to obtain useful information [39].

7.7.3   Ant Colony Optimization

Meta-heuristic algorithms such as ant-based clustering algorithm show very promising performance in data mining [16, 40]. The problem of finding the right number of clusters is considered as an NP-hard problem [16]. Therefore, meta-heuristic algorithms can be applied as clustering algorithm in solving NP-hard problem. Ant-based clustering algorithm is inspired by the real ant colonies when they cluster the corpses and larvae sorting. Below code depicts the pseudo code for ant-based clustering approach in Big data.

```
 1: Begin
 2: Initialization phase
 3: Randomly scatter all data on the grid
 4: While (termination condition not met) do
 5: Each ant randomly picks up one data item
 6: Each ant randomly placed on the grid
 7: For each ant (i = 1, . . . , n) do
 8: While (ant[i] carries item)
 9: ant[i]:= move randomly on the grid
10: if (ant[i] decide to drop item) do
11: ant[i]:= drop item
12: End while
13: End for
14: End while
15: End
```

The algorithm's basic principle focuses on agents where the agents represent the ants that randomly move around in their environment which is a squared grid with periodic boundary conditions. While ants wandering around in their environment, they pick up the data item that are either isolated or surrounded by dissimilar ones. The picked item will be transported and dropped by ants to form a group with a similar neighbourhood items base on similarity and density of data items.

The probability of picking an element increases with low density and decreases with the similarity of the element. The idea behind this type of aggregation pheromone is the attraction between data items and artificial ants. Small clusters of data items grow by attracting ants to deposit more items. Therefore, this positive feedback leads to the accumulation of larger clusters.

Clusters ensemble approach by Yang and Kamel [54] which incorporates multi-ant colonies algorithm for clustering is suggested. The approach consists of two parts. The first part focuses on several independent and heterogeneous ant colonies where each uses ant-based clustering algorithm. In the second part, a queen ant agent (also called master) aggregates the output clusters from each ant colony using a hyper graph model which is proposed by Strehl and Ghosh [49]. Each ant colony works in parallel and produced clusters and sent them to the queen ant agent. The queen ant agent combines the clusters to update and broadcast the similarity matrix and then the procedure is iterated.

Ant-based algorithm has many advantages to be used in Big data mining because it has the ability to scale with the size of the data set, prior knowledge of the number of expected clusters is not needed and easy to integrate with clusters ensemble model. Big data analysis opens the door for many research areas and one of the most important areas is the data security.

7.8   Conclusion

In this new era with boom of data both structured and unstructured, in the field of genomics, meteorology, biology, environmental research and many others, it has become difficult to turning unstructured, invaluable, imperfect, complex data into usable information from a hidden or a complex data set using traditional algorithms/methods. The very fact that Big data analysis typically involves multiple phases highlights a challenge that arises routinely in practice: production systems must run complex analytic pipelines, or workflows, at routine intervals, e.g., hourly or daily. New data must be incrementally accounted for, taking into account the results of prior analysis and pre-existing data. And of course, provenance must be preserved, and must include the phases in the analytic pipeline. Current systems offer little support for such Big data pipelines, and this is in itself a challenging objective. In this scenario, proper algorithms and methods are needed to classify the data, find a suitable pattern among them.

The chapter describes few of these challenges for handling Big data and the features of Evolutionary algorithm for which it can used as a feasible option to handle such huge data sets. And it also gives an overview of the applications of Big data where Evolutionary algorithm has been implemented as a successful solution to handle this kind of complex and large data sets.

References

1. Abraham, A., Grosan, C., & Ramos, V. (2006). *Swarm intelligence in data mining* (Vol. 34). Heidelberg: Springer.
2. Alexandros, L., & Jagadish, H. (2012). Challenges and opportunities with big data. *Proceedings of the VLDB Endowment*, *5*, 2032–2033.
3. Back, T., Hammel, U., & Schwefel, H.-P. (1997). Evolutionary computation: Comments on the history and current state. *IEEE Transactions on Evolutionary Computation*, *1*, 3–17.
4. Bandyopadhyay, S., & Muthy, C. (1995). Pattern classification using genetic algorithms. *Pattern Recognition Letters*, *16*, 801–808.
5. Bellman, R. (1961). *Adaptive control processes: A guided tour*. Princeton: Princeton University Press.
6. Bentley, P. (2000). Evolutionary, my dear Watson—Investigating committee-based evolution of fuzzy rules for the detection of suspicious insurance claims. In *Genetic and Evolutionary Computation Conference (GECCO-2000)*. New York: Morgan Kaufmann.
7. Blickle, T., & Thiele, L. (1995). A comparison of selection schemes used in genetic algorithms. Technical 11, TIK.
8. Bosman, P., & Thierens, D. (2003). The balance between proximity and diversity in multiobjective evolutionary algorithms. *IEEE Transactions on Evolutionary Computation*, *7*(2), 174–188.
9. Brockhoff, D., & Zitzler, E. (2009). Objective reduction in evolutionary multiobjective optimization: Theory and applications. *Evolutionary Computation*, *17*(2), 135–166.
10. Cantu-Paz, E., & Kamath, C. (2001). *On the use of evolutionary algorithms in data mining*, H. A. Abbass, R. A. Sarker & C. Sincla (Eds.). San Francisco, CA: Morgan Kaufmann.
11. Chen, S. (2002). *Genetic algorithms and genetic programming in computational finance*. Boston, MA: Kluwer.
12. Cheng, S., Shi, Y., Qin, Q., & Gao, S. (2013). Solution clustering analysis in brain storm optimization algorithm. In *Proceedings of The 2013 IEEE Symposium on Swarm Intelligence (SIS 2013)* (pp. 111–118). Singapore: IEEE.
13. Chui, M., Loffler, M., & Roberts, R. (2010). *The internet of things* (Vol. 2).
14. Coello, C., Dehuri, S., & Ghosh, S. (2009). *Swarm intelligence for multi-objective problems in data mining* (Vol. 242). Heidelberg: Springer.
15. Cohen, S., & de Castro, L. (2006). Data clustering with particle swarms. In *Proceedings of the 2006 IEEE Congress on Evolutionary Computations (CEC 2006)* (pp. 1792–1798).
16. Das, S., Abraham, A., & Konar, A. (2009). Metaheuristic pattern clustering an overview. *Metaheuristic clustering SE 1* (Vol. 178, pp. 1–62). Berlin, Germany: Springer.
17. Deb, K. (2001). *Multi-objective optimization using evolutionary algorithms*. UK: Wiley.
18. Desarkar, A., & Garai, G. (2014). Comparative study of traditional and modified selection operators on mathematical functions for evolutionary algorithm. In *Proceedings of International Conference on Emerging Research in Computing, Information, Communication and Applications, ERCICA-14*.
19. Dzeroski, S., & Lavrac, N. (2001). *Relational, mining, data*. Secaucus, NJ: Springer. Fayyad, U., Piatetsky-Shapiro, G., Smyth, P., & Uthurusamy, R. (1996). *Advances in knowledge discovery and data mining* (p. 2001). Melno Park, CA: The MIT Press.

20. Eiben-Smith. Evolutionary algorithm chapter 2. http://www.cs.vu.nl/~gusz/ecbook/Eiben-Smith-Intro2EC-Ch2.pdf.
21. Engelbrecht, A. (2007). *Computational intelligence: An introduction* (2nd ed.). Sussex: Wiley.
22. Forrest, S. (1993). Genetic algorithms: Principles of natural selection applied to computation. *Science, 261*, 872–878.
23. Freitas, A. A. (2003). A survey of evolutionary algorithms for data mining and knowledge. In A. Ghosh & S. Tsutsui (Eds.), *Advances in evolutionary computing: theory and applications* (pp. 819–846). New York, NY: Springer.
24. Freitas, A. A. (2002). *Data mining and knowledge discovery with evolutionary algorithms.* Berlin: Springer.
25. Gebhardt, F. (1991). Choosing among competing generalizations. *Knowledge Acquisition, 3*(4), 361–380.
26. Goldberg, D., & Deb, K. (1991). *A comparative analysis of selection schemes used in genetic algorithms.* In G. J. E. Rawlins (Ed.). Los Altos: Morgan Kaufmann
27. Gonzáler, M. C., Hidalgo, C. A., & Barabási, A. L. (2008). Understanding individual human mobility patterns. *Nature, 453*, 779–782.
28. Hand, D. M. (2001). *Principles of data mining.* Cambridge: MIT Press.
29. Hastie, T., Tibshirani, R., & Friedman, J. (2009). *The elements of statistical learning: Data mining, inference, and prediction* (2nd ed.). Springer Series in Statistics. Heidelberg: Springer.
30. Holland, J. (1962). Outline for a logical theory of adaptive systems. *Journal of the ACM, 9*(3), 297–314.
31. Jadaan, O. A., Rajamani, L., & Rao, C. R. (2005). Improved selection operator for GA. *Journal of Theoretical and Applied Information Technology, 4*(4), 269–277.
32. Jin, Y., & Branke, J. (2005). Evolutionary optimization in uncertain environments a survey. *IEEE Transactions on Evolutionary Computation, 9*(3), 303–317.
33. Joshi, A., Wallwork, J., AlYahya, K., & AlOtaibi, S. The use of evolutionary algorithms in data mining. http://www.cs.bham.ac.uk/ rjh/courses/NatureInspiredDesign/2010-11/StudentWork/Group3/The%20Use%20of%20Evolutionary%20Algorithms%20in%20Data%20Mining.pdf.
34. Julstrom, B. A. (1999). *It's all the same to me: Revisiting rank-based probabilities and tournaments.* Department of Computer Science: St. Cloud State University.
35. Kulkarni, R., & Venayagamoorthy, G. (2011). Particle swarm optimization in wireless- sensor networks: A brief survey. *IEEE Transactions on Systems, Man, and Cybernetics, Part C: Applications and Reviews, 41*(2), 262–267.
36. Kumar, R., & Jyotishree, (2012). Blending roulette wheel selection and rank selection in genetic algorithms. *International Journal of Machine Learning and Computing, 2*, 4.
37. Lu, Y., Wang, S., Li, S., & Zhou, C. (2011). Particle swarm optimizer for variable weighting in clustering high-dimensional data. *Machine Learning, 82*(1), 43–70.
38. Martens, D., Baesens, B., & Fawcett, T. (2011). Editorial survey: Swarm intelligence for data mining. *Machine Learning, 82*(1), 1–42.
39. Pal, S. K., Talwar, V., & Mitra, P. (2002). Web mining in soft computing framework: Relevance, state of the art and future directions. *IEEE Transactions on Neural Networks, 13*(5), 1163–1177.
40. Pancerz, K., Lewicki, A., & Tadeusiewicz, R. (2012). *Ant based clustering of two-class sets with well categorized objects* (Vol. 299). Berlin, Germany: Springer.
41. Pei, M., Punch, W., & Goodman, E. (1998). Feature extraction using genetic algorithms. In *Proceedings of International Symposium on Intelligent Data Engineering and Learning (IDEAL'98).*
42. Rajaraman, A., Leskovec, J., & Ullman, J. (2012). *Mining of massive datasets.* Cambridge: Cambridge University Press.
43. Razali, N. M., & Geraghty, J. (2011). Genetic algorithm performance with different selection strategies in solving TSP. In *Proceedings of the World Congress on Engineering (WCE 2011)* (Vol. 2).
44. Scott, D., & Thompson, J. (1983). Probability density estimation in higher dimensions. In J. E. Gentle (Ed.), *Computer Science and Statistics: Proceedings of the Fifteenth Symposium on the Interface.*

45. Sheppard, J., & Salzberg, S. (1997). A teaching strategy for memory-based control. *Artificial Intelligence Review, 11*, 343–370.
46. Siedlecki, W., & Sklansky, J. (1989). A note on genetic algorithms for large-scale feature selection. *Pattern Recognition Letters, 10*, 335–347.
47. Sivaraj, R., & Ravichandran, T. (2011). A review of selection methods in genetic algorithm. *International Journal of Engineering Science and Technology (IJEST), 3*(5), 3792–3797.
48. Slaney, M., & Casey, M. (2008). Locality-sensitive hashing for finding nearest neighbors. *IEEE Signal Processing Magazine, 25*(2), 128–131.
49. Strehl, A., & Ghosh, J. (2002). Cluster ensembles a knowledge reuse framework for combining multiple partitions. *Journal of Machine Learning Research, 3*, 583–617.
50. Teodorovic, D. (2003). Transport modeling by multi-agent systems: A swarm intelligence approach. *Transportation Planning and Technology, 26*(4), 289–312.
51. Vafaie, H., & Jong, K. (1994). *Improving a rule induction system using genetic algorithms.* In R. S. Michalski, R. S. Michalski & G. Tecuci (Eds.). San Francisco, CA: Morgan Kaufmann.
52. Verma, G., & Verma, V. (2012). Role and applications of genetic algorithm in data mining. *International Journal of Computer Applications, 48*, 17.
53. Walter, D. (2000). Cladia: A fuzzy classifier system for disease diagnosis. In *Proceedings of the Congress on Evolutionary Computation.*
54. Yang, Y., & Kamel, M. (2006). An aggregated clustering approach using multi-ant colonies algorithms. *Pattern Recognition, 39*(7), 1278–1289.
55. Zaki, M. J., Yu, J. X., Ravindran, B., & Pudi, V. (2010). Advances in knowledge discovery and data mining. In *Part I, Proceedings of the 14th Pacific-Asia Conference (PAKDD 2010).* New York: Springer.
56. Zhong, J., Hu, X., Gu, M., & Zhang, J. (2005). Comparison of performance between different selection strategies on simple genetic algorithms. In *Proceedings of the International Conference on Computational Intelligence for Modeling, Control and automation, and International Conference of Intelligent Agents, Web Technologies and Internet Commerce.*

Chapter 8
Statistical and Evolutionary Feature Selection Techniques Parallelized Using MapReduce Programming Model

M. Janaki Meena and S.P. Syed Ibrahim

Abstract Advances in computer technologies and internet have accumulated data at an exceptional speed. Majority of data available is in textual form and might contain billions of significant observations and thousands of features. In the era of Big data, text categorization is the primary step for handling and organizing textual data. Mining of knowledge from unstructured data is a more challenging task and proper representation of unstructured data will significantly improve the performance of the knowledge extraction process. Bag of words is a basic text representation technique, but has not been much successful due to the properties synonymy and polysemy of the English language. These problems could be reduced when documents are represented as a bag of phrases. Finding the best subset of features for a problem in a large domain becomes intractable and many such problems have been proved to be NP-Hard. Optimization algorithms are designed to approach such NP-Hard problems to find nearly optimal solutions. This chapter discusses about using MapReduce programming model for statistical feature selection when documents are represented as a bag of syntactic phrases and designing a parallel ant colony optimization algorithm for feature selection.

Keywords Big data · Text classification · Hadoop · Feature selection

8.1 Introduction

Enormous amount of textual data is generated every day by corporations, web pages, emails, reports in stock markets, social group networks, newsgroups etc. Using the generated Big data, a number of intelligent information systems such as product categorization, search personalization, product reviews classification, sentiment mining, document filtering for digital libraries, automatic email filtering and biographic gen-

M. Janaki Meena (✉) · S. Syed Ibrahim
SCSE, VIT University Chennai Campus, Chennai 600127, India
e-mail: janakimeena.m@vit.ac.in

S.P. Syed Ibrahim
e-mail: syedibrahim.sp@vit.ac.in

© Springer International Publishing Switzerland 2016   159
B.S.P. Mishra et al. (eds.), *Techniques and Environments for Big Data Analysis*,
Studies in Big Data 17, DOI 10.1007/978-3-319-27520-8_8

eration are built. Categorization of documents is an important step in building such intelligent systems and the features included for the categorization process makes a difference in the performance of the classifier.

As textual data is in unstructured form, the first step in the classification process is to find a better representation of the documents to feed them into the classifier. Text representation is a vital step in classification as it tries to capture the meaning of documents, and help the learning process [10]. For text mining tasks, documents may be represented as a 'bag of words' (BOW). It is a numeric vector representation in which each element corresponds to the count of occurrence of a term in the document [17]. BOW is not an ideal representation as it ignores the position of the words and its context. The properties such as synonymy and polysemy of English lead to ambiguity which in turn weakens the performance of the machine learning algorithms when documents are represented as BOW [12].

Ambiguous meaning of words gets reduced when documents are represented as phrases, co-occurrences of words in the documents. Documents may be represented by two types of phrases, syntactic phrases and statistical phrases.

N-gram syntactic phrase is the occurrence of 'n' words in a document with certain syntactic relationship. Examples of noun phrases are 'the red car that crossed the signal', 'the boy with a cup', and 'the running machine'. And examples of verb phrases are 'playing cricket', and 'come to the shop'. A n-gram statistical phrase is a sequence of n-words that occur frequently. From literature, it is observed that statistical phrases are inferior to syntactical phrases due to the inferior statistical qualities of phrases and the fact that the same concept may be elicited by linguistically different but related units. It has also been observed that phrases of size two or three aids the machine learning process while inclusion of longer phrases decreases the performance.

As text mining tasks are high dimensional problems and the dimensionality of the problem still becomes larger by including phrases for document representation, feature selection becomes an important preprocessing step. If feature selection is done poorly, no clever learning algorithm can compensate, and when it is done well, the computational and memory demands of both the inducer and the predictor can be reduced [9].

This chapter designs an algorithm to identify syntactic bigrams and filter them using a statistical technique based on CHI algorithm. POS tagging is done using Wordnet and basic phrase structure rules. Though a number of statistical algorithms had been proposed for feature selection, the problem has also been observed as a combinatorial problem with an objective function to minimize the error of the classifier. Combinatorial problems could be solved by exhaustive search, a brute and force technique. But for a larger problem domain, it is impossible to perform exhaustive search and that makes the problem to be NP hard.

Ant Colony Optimization is a metaheuristic technique to find nearly optimal solutions for combinatorial optimization problems. Metaheuristic methods are designed to guide a problem specific heuristic towards the high quality solution regions. ACO models have characteristics such as distributed computation, positive feedback, and greedy heuristics [8].

This chapter also designs an optimization algorithm to identify the subset of features that have better classification accuracy. The search space for the ACO problem is formulated as a graph with every word as a node and all words are connected to each other. The edges of the graph indicate the order of visit of the nodes as any word can be chosen after any other word in the search space, the graph is constructed as a fully connected graph. Every node in the graph is associated with a heuristic information and a pheromone value.

Detecting syntactic phrases is a complex process and the time taken for the execution of the algorithm is very high with a large training corpus. The proposed ACO algorithm is a wrapper model, hence it takes a longer time to complete one iteration in addition to taking a number of iterations to find a good solution. Both the algorithms discussed in this chapter are time intensive. Hence the algorithm are parallelized based on MapReduce programming model and executed in a cluster of machines. The effectiveness of the features selected by the proposed methods was analyzed with the nave bayes classifier. Nave bayes classifier was chosen for experimentation for two reasons. The first one is its simplicity and sensitiveness to the features fed into it [15]. The second one is that ambiguity in features introduces noise in the nave bayes classifier.

8.2   Concepts Involved in the Research

This section of the chapter discusses about the concepts involved in solving the problem.

8.2.1   Feature Selection Techniques

Any feature selection algorithm consists of four steps such as subset generation, subset evaluation, stopping criterion, and result validation [4]. The feature space is searched to generate candidate feature subsets and an evaluation criterion is formulated to evaluate the generated subsets. This process is iteratively done till the stopping criterion is reached.

Feature selection algorithms are classified into filter model, wrapper model and hybrid model based on the evaluation criteria used. Filter model works in general, independent of the classifier to determine the appropriate subset of features. This method evaluates the selected subset using mathematical techniques that assess the general characteristics of the data. Wrapper model asses the selected subset of features by using the mining algorithm for which the feature selection is done. Wrapper model give better performance when compared to filter model but it is computationally more expensive. Hybrid model tries to take the advantage of the two models, better accuracy and lesser complexity. This method utilizes different evaluation criteria in different search stages. The problem of feature selection becomes intractable for larger input domains and has been shown to be NP-hard [14].

8.2.2   Combinatorial Optimization Problems

Combinatorial optimization problem is similar to subset generation problem which has a number of solutions and each solution is associated with a cost. Solutions for a combinatorial problem require a methodology to generate a permutation, a combination, or a subset that satisfies certain constraints. As the first step, the problem is formulated with a set of discrete variable and an objective function. Optimal solution for the problem is found with respect to the given objective function. Solution to such a problem is found by assigning values for the discrete variables, such that the objective function is maximized or minimized. In general, a combinatorial optimization problem is defined using a triplet $\prod = (S, f, \Omega)$, where S is the set of candidate solutions, f is the objective function which assigns a numerical value to each solution in S and Ω is a set of constraints. Set of solutions $\bar{S} \subseteq S$ that satisfy the constraints Ω are called feasible solutions. The goal is to find a globally optimal feasible solution $S^* \subseteq \bar{S}$. Combinatorial problems can be solved by two categories of algorithms namely exact and approximate. Exact algorithms find the optimal solution for the problem but the computation time required to find such a solution is extremely high. Exhaustive search is an example for exact algorithm. But the number of possible solutions increases exponentially with the instance size hence becomes infeasible for problems of high dimensionality.

Approximate algorithms search for nearly optimal solutions at a relatively low computational cost. Two types of approximate techniques namely constructive and local search are proposed for combinatorial optimization problems. Constructive methods build the solution in an incremental way by adding solution components. Local search iteratively explore the neighborhoods of the current solution by local changes [8].

8.2.3   Syntactic Phrases

The proposed system aims at representing documents through strong unigrams and bigrams. BOW representation destructs the semantic relations between words. Even stable phrases, such as Information Retrieval, are separated as individual words and their meaning is lost when documents are represented as BOW. This technique aim at extracting syntactic bigrams to supplement unigrams in categorization. Consider the phrases (a)–(i) they are semantically related but linguistically different.

(a) algorithmic analysis—NP
(b) analysis of algorithm—NP
(c) analyser of the algorithm—NP
(d) algorithm analysis—NP
(e) algorithm and its analysis—NP
(f) analysing the algorithm—VP
(g) analysed the algorithm—VP

(**h**) algorithm was analysed—VP
(**i**) algorithms were analysed—VP

These phrases are different due to their morphosyntactic variations. Morphosyntactic variation of phrases include different forms of noun phrases, verb phrases and full sentences. When these phrases are preprocessed by stop word removal and stemming, they become equivalent though they give different meanings. The proposed method handles morphosyntactic variations of phrases by considering them as occurrences of the same n-gram. Morphosyntaticlly varied noun phrases are shown in (a)–(e) and verb phrases are shown in (f)–(i).

The n-grams definition is based on the hypothesis that various syntactic expressions may convey the same concept, and hence they must be seen as a form of conflation. This type of generalization performed by means of n-grams has its problems such as over-generalization and under-generalization. The sample phrase (c) is over-generalized since it does not refer to the same concept as the other examples. Under-generalization can be seen in the phrase analysis of a good algorithm; arguably it refers to the same concept as examples (b), (h) and (i) but is not recognized as such. To avoid the problem of under generalization the proposed algorithm analyzes phrases by ignoring adjectives. POS tagging is done using basic syntactic rules of English and Wordnet.

8.3 Ant Algorithms and Its Types

Ant algorithms are metaheuristic algorithms that intend to determine approximate solution for optimization problems. The idea of these ant algorithms came by the observations made in the foraging behaviour of ants and stigmergy. Stigmergy refers to the indirect communication among individuals modifying their local environment in a self-organizing emergent system. Deneubourg et al. [5], investigated the strigmeric nature of ant colonies by conducting a number of experiments. Their experiments showed that ants communicate each other by laying down pheromone trails and their path converges in due course of time as ants have the tendency to follow trails with higher pheromone concentration.

Artificial algorithms are built using stochastic model built based on the behavior observed from natural ants. In the ACO algorithm, a search space is created with nodes (referred to as states) and a procedure is designed to find a solution path. ACO algorithms generate artificial ants and determine solution for the problems by working in an iterative fashion. The artificial ants communicate to each other indirectly via a synthetic pheromone. All the ants generated to solve the problem act like a colony and each ant is a simple computational agents that work cooperatively and communicate each other by artificial pheromone trails. The movement of the ants in the search space is guided by two factors of each node in the search space:

1. Heuristic information: a factor specific to the problem and is determined a priori to the run of the algorithm. It is a measure of preference for an ant to move from a node or state Si to a node or state Sj.
2. Artificial Pheromone Trails: This factor represents the quality of the prior solutions when that node was included in the solution path. This factor is updated upon completion of each iteration.

The main properties of artificial ants are [3]:

- Each artificial ant has an internal memory to store the partial solution determined by it.
- Each ant starts in a random initial state $S_{initial}$ of the search environment and moves in an iterative fashion.
- Each movement of the ant is decided based on a transition rule which contains the guidance factors heuristic information and pheromone of the nodes.
- Each ant deposit certain amount of pheromone either on the states or on the state transition after every iteration. The quantity of pheromone to be deposited is determined by a problem specific pheromone update rule.
- The solution determined by the artificial ants could be improved by a local search [7].
- ACO algorithms could be speeded up for convergence by designing daemon actions such as depositing additional pheromone on the states of the global best solution.

Dorigo et al. [7] developed the first ant algorithm for traveling salesman problem. They considered each city as a state (node) and the path connecting the cities as transitions (edges). Pheromone is deposited either on the states or on the edges to indicate the importance of including them in the solution. In the first ant algorithm for TSP, pheromone was deposited on the connected paths. Heuristic function for TSP was defined as inverse of the distance between the towns. In each step of the algorithm, one among the randomly chosen 'm' number of cities was included in the tour. A probabilistic rule known as the transition rule is used to determine the city that is to be included in the tour. Transition rule was defined using the heuristic information and pheromone deposits on the node. The algorithm maintained a tabu list to store the transitions of already visited towns which was used to maintain the constraint of the problem. Each iteration of the problem determined a tour. After each tour, based on the strength of the determined solution, ants lay pheromone along each visited path.

Dorigo et al. [6] modified the ant algorithm and proposed ACO. ACO algorithm included the concept of pheromone evaporation in addition to the construct solution and pheromone update modules of ant systems. The concept of pheromone evaporation was introduced to make the ants to forget bad solutions. Stutzle et al. [18] improvised ACO algorithm and termed it as Max-Min Ant System Algorithm. It differs from earlier ACO algorithm in two major ways: global best ant only updates pheromone and the pheromone update function is bounded by limits. The algorithm define a lower and upper limit for the pheromone value. Few more variations of ant algorithms evolved in 1990s. Bullnheimer et al. [2] developed a rank based ant system in which the ants were arranged in a sorted order based on the quality of their solutions. Pheromones were updated for the nodes in the solutions found by the first 'm' ants and the amount of pheromone deposited was directly proportional to the quality of solution. Dorigo and Gembardella [7] varied the pheromone update function. They employed a local pheromone update in addition to the pheromone update at the end of each epoch. Experimental results of Stutzle et al. [18] showed that min-max ant system performs better than the other ant colony algorithms for TSP.

8.4 CHI Method

This is a statistical feature selection method that ranks terms based on their dependencies to categories. Terms that have a strong dependency are selected as features. CHI(χ^2) value for a term is obtained by comparing the observed co-occurrence of frequencies in a 2 way contingency table with the expected frequencies when they are assumed to be independent. In a corpus of 'm' categories, that contains 'n' labeled documents in total, a 2×2 contingency table is formed for each term with respect to each category as shown in Table 8.1.

Expected frequency E (i, j), where i represent the presence or absence of a feature and j represents whether the document belongs to a category is calculated using (8.1) [13]:

$$E(i, j) = \frac{\sum_{a \in \{w, \neg w\}} O(a, j) \sum_{b \in \{c, \neg c\}} O(i, b)}{n} \tag{8.1}$$

The χ^2 statistics value for the term 'w' with respect to the category 'c' is defined as (8.2):

$$\chi^2_{w,c} = \sum_{i \in \{w, \neg w\}} \sum_{j \in \{c, \neg c\}} \frac{(O(i, j) - E(i, j))^2}{E(i, j)} \tag{8.2}$$

The degrees of freedom for a contingency table of dimension r, c is $(r - 1) \times (c - 1)$. Hence in this example, the degree of freedom is $(2 - 1) \times (2 - 1) = 1$. The value obtained by using Eq. (8.2) is compared with the value in the standard χ^2 tabulation for the determined degrees of freedom with confidence level 0.1 %. The value obtained from table for the degrees of freedom 1 with 0.1 % confidence level is 10.83. χ^2 value of a term 'w' which has the occurrence as shown in Table 8.1 is 172.

Table 8.1 A 2×2 term-category contingency table for term 'w' with respect to category 'c'

	c	¬c	Σ
w	90	35	125
¬w	50	350	400
Σ	140	385	525

Table 8.2 A 2×2 Term-Category Contingency Table for term 'w_1' with respect to category 'c'

	c	$\neg$c	Σ
w_1	50	350	400
$\neg w_1$	90	35	125
Σ	140	385	525

Goodness of fit is used to determine if there is a dependency between a term and category. Measures of goodness of fit find out whether there is a discrepancy between observed value and the value expected under the model in question. For a corpus with 'm' classes, term-goodness is defined as either the average as defined in (8.3) or the maximum as defined in (8.4) [13]:

$$\chi^2_{avg}(w) = \sum_{j=1}^{m} p(c_j)\chi^2_{w,c_j} \tag{8.3}$$

$$\chi^2_{max}(w) = max_j\{\chi^2_{w,c_j}\} \tag{8.4}$$

$p(c_j)$ is the probability of the documents to be in the category c_j. CHI algorithm determines only the if there is a dependency between a term and a category. It does not concern about the type of dependency [13].

The χ^2 value of another term 'w_1' which has the occurrence as inverse as that of 'w' as depicted in Table 8.2 is also 172. It may be observed from Table 8.1 that there is positive dependency between 'w' and 'c' since $90/140 \approx 2/3$ of the documents in 'c' contain 'w' and $90/125 \approx 3/4$ of the documents containing 'w' are in 'c'. Whereas it is not clear whether the term 'w_1' has a positive dependency with 'c', since $50/140 \approx 1/3$ of the documents in 'c' only contain 'w' and only $50/400 \approx 1/8$ of the documents containing 'w' are in 'c'.

8.5 CHIR Method

CHIR is an extended technique based on χ^2 statistics and it was proposed by Li et al. [13]. This technique find out the type of dependency between a term and a category as well.

To determine whether the dependency between 'w' and 'c' is positive or negative, a new measure $R_{w,c}$ was defined in CHIR as (8.5):

$$R_{w,c} = \frac{O(w, c)}{E(w, c)} \tag{8.5}$$

The value of $R_{w,c}$ is close to 1, when there is no dependency between the term w and the category c. When there is a positive dependency between the term and the category, the observed frequency is larger than the expected frequency. Hence the value of $R_{w,c}$ is larger than 1. For a word 'w' that has negative dependency to a category 'c', $R_{w,c}$ is smaller than 1. Based on χ^2 statistics and $R_{w,c}$ a new definition for term-goodness for a corpus with m classes is given in CHIR algorithm as shown in (8.6):

$$r\chi^2(w) = \sum_{j=1}^{m} p(R_{w,c_j})\chi^2_{w,c_j} \ with \ R_{w,c_j} > 1 \tag{8.6}$$

where $p(R_{w,c_j})$ is the weight of χ^2_{w,c_j} in the corpus. In terms of R_{w,c_j}, $p(R_{w,c_j})$ is defined as:

$$p(R_{w,c_j}) = \frac{R_{w,c_j}}{\sum_{j=1}^{m} R_{w,c_j}} \ with \ R_{w,c_j} > 1 \tag{8.7}$$

Larger values of $r\chi^2(w)$ indicates that the term w is more relevant to the category.

8.6   Distributed and Parallel Computation Using Hadoop

Enormous volume of data is being generated by both machine and human and most of the data is also made public every year. Data are generated automatically from machine logs, RFID readers, sensor networks, GPS vehicle traces, and etc. To compete in the growing knowledge economy, industries have to find an effective way to manage their own data and external data. Though storage capacities of hard drives and the number of instructions executed per second by the processor have increased tremendously, the reading and writing rate of data has not improved much. The amount of data to be analyzed has grown larger than the storage and processing capacity of a single machine. Even highly efficient algorithms may be beaten by Big data. One of the possible way to approach this problem is to distribute the data among many machines and make them to work in parallel.

The challenges that are to be addressed in distributed computing are hardware failure and combining results of individual execution. The most popular tool available to process Big data in a distributed environment is Hadoop. Using Hadoop for processing Big data is also cost effective as the theoretical 1000-CPU machine would cost a very large amount of money, when compared to a cluster of 1,000 single-CPU or 250 quad-core machines. Hadoop is a distributed batch processing infrastructure to process data in the order of hundreds of gigabytes to terabytes or even petabytes. It ties small and reasonably priced machines together into a single cost effective cluster. As commodity machines are connected to process the data, the first important challenge to be addressed is to handle hardware failure. Hadoop handles this challenges by replicating data across the cluster. Hadoop has designed a special file system Hadoop Distributed File System (HDFS). When data is submitted to the HDFS it is divided into blocks and each block is replicated as per the replication factor specified in the configuration of the cluster. The next challenge associated with distributed processing is to combine output data from different machines. The Hadoop framework provides support to combine data by the MapReduce programming model and the shuffle and sort module [19].

8.6.1 Hadoop Distributed File System (HDFS)

HDFS is a distributed file system designed based on the Google File System (GFS) to hold very large amounts of data. Individual files are broken into blocks of a fixed size and stored across the cluster. Each machine in the Hadoop cluster is referred to as a DataNode. By default, block size in HDFS is 64 MB. Block size in HDFS is made larger so that the amount of metadata storage required per file is less. Further, larger blocks allow fast streaming reads of data, as large amounts of data is sequentially laid out on the disk. To store its metadata reliably, it is all handled by a single machine, called the NameNode. The NameNode keeps track of only the file names, permissions, and the locations of each block of each file in the file system. Because of the relatively low amount of metadata per file, all of this information can be stored in the main memory of the NameNode machine, allowing faster access to the metadata. Reliability of data is an very important essential property of any file system. As commodity machines are connected to form a cluster of machines, to manage hardware failure, replica of data is maintained in the cluster. By default, HDFS maintains three copies of each block of data and the number of replicas can be configured as required by the user.

8.6.2 MapReduce Programming Model

MapReduce programming model is a batch query processor, and has the ability to run an ad hoc query against the whole dataset to get the results in a reasonable time. This programming model in Hadoop is developed based on MapReduce programming model of Google. It is designed to run jobs those last for minutes to hours on trusted, dedicated hardware running. All data elements in MapReduce are immutable, meaning that they cannot be updated. Hadoop can run MapReduce programs written in various languages such as Java, Ruby, Python, and C++.

The basic working principle of MapReduce is to break a problem into independent pieces to be worked on in parallel. MapReduce splits the processing into two phases: the map phase and the reduce phase. Each phase has key-value pairs as input and output, the type of which may be chosen by the programmer. The programmer also specifies two functions: the map function and the reduce function. The map function takes an input key/value pair and produces a set of intermediate key/value pairs. The MapReduce framework groups together all intermediate values associated with the same intermediate key and passes them to the reduce function. The intermediate values are passed to the users' reduce function through an iterator. The reduce function, written by the user takes the intermediate key and a set of values for that key as input. The input and output of the Map and Reduce function is given as below:

$$map\ (k1, v1) \rightarrow\ list(k2, v2)$$
$$reduce\ (k2,\ list(v2)) \rightarrow\ list(v2)$$

The MapReduce framework automatically parallelizes the code and runs in it a cluster of machines. One of the machines in the cluster acts as the master and the rest of the machines are slaves. The overall flow of a MapReduce operation goes as follows.

1. The MapReduce library splits the input file into 'M' pieces and start up several copies of the map module on 'M' machines of the cluster with 'N' machines. Here 'N' is greater than 'M'.
2. One of the machines in the cluster acts as the master, and the rest of the machines are slaves controlled by the master. The number of map tasks and reduce tasks may be set by the user both statically before the job is started or during the execution of the job. The job is submitted to the master. The master picks up the lightly loaded slave machines and assigns each one them a map or a reduce task.
3. Each mapper machine reads the content of the input-split assigned to them as a key value pair. A number of input formats are defined in Hadoop to process data. By default, TextInputFormat is used for reading input files. In this format, the input file is read line by line, key is the byte offset of the line in a file and value as the content of the line. The user defined Map function work on the input key value pair and emit an output key value pair. These are intermediate data generated that are buffered in the memory.
4. Periodically, the buffered pairs are written to the local disk by the Hadoop framework and the locations of these buffered data are passed on to the master. The partitioning function splits the buffered data into 'R' regions. The partitioner can be configured programmatically to control the separation of data. Then, the master communicates the locations of data to the reduce workers.
5. The reducers uses remote procedure calls to read the buffered data from local disks of mappers. When a reducer has read all intermediate data, it sorts and group data by intermediate keys. So that all occurrences of the same key are grouped together.
6. For each intermediate key, the reducer passes the key and the corresponding set of intermediate values to the reduce function. Reduce function is defined by the programmer to process a key and a list of values corresponding to it. The output of the reduce function which is also a key value pair is appended to a final output file maintained for this reduce partition. By default the output format is TextOutputFormat in which the key value pair is written to the output file separated by a tab.
7. When the entire map and reduce tasks have been completed, the master wakes up the user program and the MapReduce call returns back to the user code.

8.6.2.1 Merits of MapReduce

High Performance Computing (HPC) and Grid Computing communities had been working on large-scale data for years, using APIs such as Message Passing Interface (MPI). In HPC the work was distributed across a cluster of machines, which access a shared file system, hosted by a SAN. Data has to be transmitted over the network to the machines where it has to be process. Therefore HPC worked well for predominantly compute-intensive jobs, but becomes a problem when nodes need to access larger data volumes, since the network bandwidth is the bottleneck and compute nodes became idle. MPI is not suitable for processing Big data.

MapReduce tries to move the code over the network to the nodes where data is present; this feature is known as data locality. As code is very small in size this programming model can successfully process Big data. This property of data locality is the heart of MapReduce and is the reason for its good performance. One of the machine in the cluster acts as the master for the jobs submitted to it and it is referred as Jobtracker and other machines in the cluster are termed as task trackers. If suppose a tasktracker fails while executing a job, the jobtracker identifies this and allocate the part of the incomplete job to another machine in the cluster. The jobtracker keep track of the healthy machines every second to identify hardware failure and handle the situation.

8.7 CHIR Algorithm for Syntactic Phrases

The procedure to select phrases using statistical method is designed as an iterative MapReduce algorithm. Part of speech tagging is the key step in the algorithm. Word-Net is used to determine the POS of each word. There are two stages in the algorithm:
1. The first stage of algorithm is to identify the valid unigrams and bigrams after stop word removal.
2. The second stage of the algorithm computes the relevancy and χ^2 value for each term (Fig. 8.1).

Each stage is implemented as a MapReduce job, output of the first MapReduce job is given as input for the second job.

MapReduce Algorithm for Job1

Input: Set of input documents
Output: Unigrams and bigrams with their category and count
Step 1: The input documents are read by MapReduce job1 using TextInputFormat.
Step 2: The MapReduce framework splits the input file into a number of parts, as defined by the split size.
Step 3: The map function is called for each line in the input file. Map function execute Step 4 to Step 6.

Fig. 8.1 System architecture for parallel phrase extraction algorithm

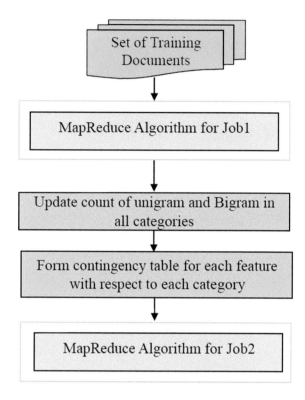

Step 4: Tokenize the line read by the mapper. For each word in the ith position, W_i of the line, Step 5 to 6.

Step 5: Discard W_i if it is a stop word.

Step 6: Otherwise connect to WordNet and find POS for W_i and W_{i+1}, if W_i and W_{i+1} are syntactically related then form a phrase with W_i and W_{i+1} in lexicographic order separated by '_'. Emit the phrase and category in which it occurs as output key of mapper and one as output value.

Step 7: One reducer is defined for each category and a partitioner function is defined to direct the extracted unigrams and bigrams to the corresponding reducer.

Step 8: Shuffle and Sort module of Hadoop framework group and sort by key data for each reducer.

Step 9: Each reducer execute Step 10.

Step 10: Sum up the value of each unigram and bigram and write the unigram or the bigram as output key and count as output value in the output file.

MapReduce Algorithm for Job2

Input: A file with the extracted unigrams and bigrams and their contingency tables
Output: A file with unigrams and bigrams and their goodness of fit, sorted as per the value of goodness of fit.

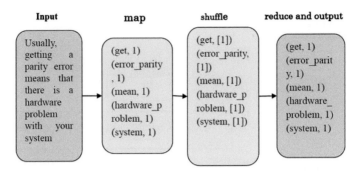

Fig. 8.2 MapReduce logical data flow for phrases extraction

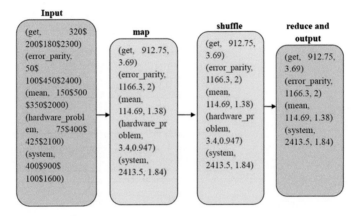

Fig. 8.3 MapReduce logical data flow for applying CHIR

Step 1: Read the input file using TextInputFormat.
Step 2: Execute Step 3 to Step 5 by mapper.
Step 3: Represent the contingency table for the term using arrays.
Step 4: Calculate goodness of fit for each term using Eqs. (8.3)–(8.7).
Step 5: Goodness of fit for the term is emitted as output key and the term is emitted as output value.
Step 6: Shuffle and sort module of the framework sort by key.
Step 7: Reducer has no role in the job, hence the number of reducers can be set as zero for this job.

Figure 8.2 represents the logical flow diagram for the MapReduce job that extract phrases from the input document. The input for the second mapper is the set of phrases and a contingency table for each phrase with respect to a category. Contingency table is given as string in which the values are entered in a row major ordering and they are separated by a '$' symbol. The mapper parses the string corresponding to the contingency table and computes χ^2 and $R_{w,c}$ value. The reduce function eliminates features with negative relevancy and output only phrases with positive relationship with a category. Figure 8.3 shows the MapReduce logical data flow for applying CHIR algorithm for bag-of-phrases representation.

8.8 ACO Algorithm for Feature Selection

As wrapper model of feature selection algorithms give better performance than filter model, the ACO algorithm designed here is a wrapper model. To design an ACO algorithm to solve feature selection problem, it is formulated as a combinatorial problem with an objective function to minimize the root mean square error of the classifier. Two constraints are defined for the problem and they are, selected features must be distinct and the number of terms selected for each category must not exceed 'F', a predefined value. Each word is represented as a node of the graph. Since any term can be included after any other term in the feature selection process, edges are provided between all the nodes and the graph becomes a completely connected graph.

The solution space of each ant is initially empty and is expanded by adding a solution component at every probabilistic decision. The sequence of decisions made by the ant in the search may be considered as a path of an ant in the decision graph [8, 11]. At the end of every iteration, the iteration best ant is made to deposit pheromone on the nodes chosen by it. Pheromone content of the nodes in the search space evaporates with time. The iterative process continues till the stopping criterion is reached. The stopping criterion may be either a number of iterations or a solution of desired quality [1, 8]. The design of the general ACO algorithm is as follows:
Repeat
for k ants
for f features
Generate 'b' random numbers that corresponds to the index of the nodes in the search space and apply probabilistic rule for them
Add the node with highest probabilistic value into the solution space of the ant
end-for
end-for
evaluate solutions generated by all ants
identify local best and global best ant
for every node n ∈ solution space of global or iteration best ant
deposit pheromone on node 'n'
end-for
for every node 'n' ∈ search space
decrease pheromone content by evaporation rate
end-for
until stopping criterion is reached

The important terms to be defined for an ACO algorithm are the heuristic and pheromone value. This algorithm uses $R_{w,c}$ defined in CHIR as heuristic value and it represents the dependency of the term with any one of the category in the training set. Heuristic function used in ACO is static and does not change during the execution of the algorithm. $R_{w,c}$ defined in (8.5) may be rewritten as (8.8) using (8.1):

$$R_{w,c} = \frac{O(w,c)n}{(O(w,c) + O(\neg w,c))(O(w,c) + O(w,\neg c))} \tag{8.8}$$

The value of n, the total number of documents in the training corpus and $(O(w,c) + O(\neg w,c))$ which is equal to the total number of documents in the category c are same for all the terms. Hence the value $\dfrac{n}{(O(w,c) + O(\neg w,c))}$ is a constant, therefore Eq. (8.8) may be rewritten as

$$R_{w,c} = \frac{O(w,c)}{O(w,c) + O(w,\neg c)} \times C_0 \tag{8.9}$$

The value of C_0 will be always greater than 1. When the term 'w' is a positive feature for a category 'c', the value of O(w,c) will be extremely larger than O(w,¬c) and $R_{w,c}$ will be greater than 1. Therefore it may be concluded that greater the value of $R_{w,c}$ for a positive feature, stronger is the dependency between w and c. However for a negative feature 'w' of a category 'c', the value of O(w,c) is very small when compared to O(w,¬c), and the value of $R_{w,c}$ is less than 1. Therefore lesser the value of $R_{w,c}$ for a negative feature, stronger is the dependency between 'w' and 'c'. As heuristic value of a node must indicate the strength of dependency between the term represented by the node and a category, it is defined to be equal to $R_{w,c}$ for positive features and as inverse of $R_{w,c}$ for negative features.

The probability that ant 'k' will include node 'i' at the time instance 't' is given as (8.10) where J^k is the random selection set of the ant [1]. η denotes the heuristic information associated with the ith node and $\tau_i(t)$ denotes the pheromone content of the node at the time instance t. α and β denotes the importance of pheromone content and heuristic information for the problem and those two parameters decide the exploration and exploitation properties of the algorithm.

$$P_k^i(t) = \begin{cases} \dfrac{(\tau_i(t))^\alpha \cdot (\eta_i)^\beta}{\sum\limits_{u \in J^k} (\tau_i(t))^\alpha \cdot (\eta_i)^\beta} & \forall i \in J^k \\ 0 & \text{otherwise} \end{cases} \tag{8.10}$$

ACO algorithm may be designed such that either the iteration best ant or the global best ant or both of them deposits pheromone on the nodes selected by them. The pheromone content of all nodes at a time instance t + 1 is given by (8.11) [1].

$$\tau_i(t+1) = (1-\rho)\tau_i(t) + \Delta\tau_i^k(t) + \Delta\tau_i^g(t) \tag{8.11}$$

Fig. 8.4 MapReduce
modules in ACO algorithm

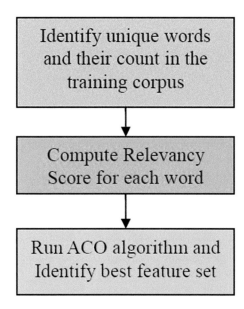

where ρ is the evaporation rate, $\Delta\tau_i^k(t)$ is the content of pheromone to be deposited by the iteration best ant k at the time instance t over the node i and $\Delta\tau_i^g(t)$ is the content of pheromone to be deposited by the global best ant g upto the time instance t, over the node i.

8.8.1 Parallelization of ACO Using MapReduce Programming Model

The proposed ACO algorithm is parallelized based on the data parallelism concept. When the algorithm was analyzed it was found that three sections of the algorithm can run in parallel. Figure 8.4 gives the details of MapReduce jobs that are designed to achieve parallelization. The first MapReduce job of the algorithm identifies the set of unique terms and their count in each category. Then a part of sequential code is required to build the contingency table for each term with respect to each category. The second MapReduce job, find the statistical heuristic information $R_{w,c}$ for each term. The third MapReduce job initiates the iterative part of the algorithm that generate the ants and the ants make a random walk in parallel to find their own solution space [16].

MapReduce Algorithm for Job1 in ACO

Input: Set of input documents

Output: Words with their category and count

Step 1: The input documents are read by MapReduce job1 using TextInputFormat.

Step 2: The MapReduce framework splits the input file into a number of parts, as defined by the split size.

Step 3: The map function is called for each line in the input file. Map function execute Step 4 to Step 6.

Step 4: Tokenize the line read by the mapper. For each word in the ith position, W_i of the line, repeat Step 5 to Step 6.

Step 5: Discard W_i if it is a stop word.

Step 6: Otherwise emit the word and category in which it occurs as output key of mapper and one as output value.

Step 7: One reducer is defined for each category and a partitioner function is defined to direct the extracted words to the corresponding reducer.

Step 8: Shuffle and Sort module of Hadoop framework group and sort by key data for each reducer.

Step 9: Each reducer execute Step 10.

Step 10: Sum up the value of each word in that category and write the word as output key and count as output value in the output file.

Output of each reducer gives the count of the word in each of the category. A sequential algorithm is formulated to read the output files and to build the contingency table.

The MapReduce logical data flow for job1 of ACO that extract unique words and their count from the training corpus is shown in Fig. 8.5.

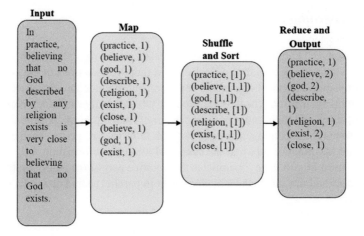

Fig. 8.5 MapReduce logical data flow to count unique words

MapReduce Algorithm for Job2 in ACO

Input: A file with the unique words in the training set and their contingency tables
Output: A file with words and their heuristic information $R_{w,c}$.
Step 1: Read the input file using TextInputFormat.
Step 2: Step 3 to Step 5 is executed by mapper.
Step 3: Represent the contingency table for the term using arrays.
Step 4: Calculate $R_{w,c}$ for each term using Eq. (8.9).
Step 5: Assign heuristic information for positive dependency terms as $R_{w,c}$ and for negative dependency term as inverse of $R_{w,c}$.
Step 6: Emit the term as output key and heuristic information as output value.
Step 7: Reducer has no role in the job, hence the number of reducers can be set as zero for this job.
The second MapReduce job reads the file with the word and its contingency table and compute $R_{w,c}$ values.

   The third MapReduce task in the ACO algorithm is computation intensive and the data required for this task is prepared by the previous MapReduce algorithms. The output file of Job2 of ACO is read in a sequential way to prepare the search space for the ants. This part of the algorithm works in an iterative fashion. Each iteration creates an MapReduce job and output of previous iteration is given as input for the current iteration. Map function does the work of an ant in ACO. For each call of Map function in every iteration an ant is simulated by the MapReduce job.

MapReduce Algorithm for Job3 in ACO

Input: Feature space initialized with heuristic information and pheromone value
Output: 'F' features selected by ACO algorithm
Step 1: Repeat Steps 2–12 for 'n' iterations
Step 2: Input file is prepared with random content in 'n' lines where 'n' is the number of ants to be generated in each iteration.
Step 3: For each line in the input file map function is called that executes Steps 4–7
Step 4: Repeat Steps 5–7 for selecting 'F' features.
Step 5: Generate 'b' random numbers, these numbers correspond to the position of the feature in the feature space.
Step 6: Apply probabilistic rule given in (8.10) for all 'b' nodes represented by random numbers
Step 7: Find maximum of the probabilistic value and include that feature into the current ant's feature space if it is not in the ant's solution space. If the feature with maximum value is in solution space of ant then include the feature with next maximum value.
Step 8: Build the Naive Bayes classifier model with the features selected by the current ant. Weka can be used for building the classifier.
Step 9: RMSE of the classifier with the features selected by the current ant is determined.
Step 10: Output key of the mapper is a common value for all ant like one and output

value is the RMSE value and the index of features separated by a delimiter.

Step 11: The shuffle and sort framework of the Hadoop framework group and sort by key. Here key is same for output of all ants therefore all values are grouped together.

Step 12: The reducer find the minimum of RMSE value (that is the best ant) and update pheromone of features selected by the iteration best ant (Fig. 8.6).

Figure 8.7 shows the MapReduce logical data flow for ACO algorithm. The input is a file with some model text. The map function ignores the text read by it and performs the operation of an ant.

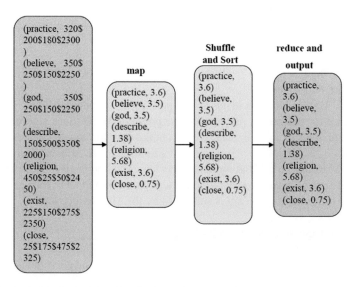

Fig. 8.6 MapReduce logical data flow for determining $R_{w,c}$ for words

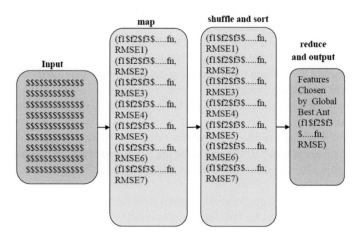

Fig. 8.7 MapReduce logical data flow for ACO algorithm

8.9   Conclusion

Every day a tremendous amount of data is generated by both human and machine. Most of the information are textual and they contain knowledge that can be applied to improve business. Dimensionality of textual data is so large that the text mining tasks cannot be effectively done with all the terms and the efficiency of text mining tasks can improve when feature selection algorithms are applied. This chapter discuss about representing documents using syntactic phrases and feature selection algorithm based on ACO. ACO is an iterative algorithm and in syntatic phrase extraction, WordNet was used to identify POS of the words so the time taken to execute the algorithms were very much high. Therefore the algorithms were parallelized using MapReduce programming model of Hadoop. The time taken to execute the algorithms decreased in proportion to the number of machines in the Hadoop cluster. It was observed that the performance of the classifier increased from 2 to 13 % for different datasets when documents were represented as a bag of phrases. ACO algorithm was executed for 150 iterations and it was found that the number of positive features selected increased with the number of iterations. The performance of the classifier increased from 2 to 5 % when the features selected by ACO algorithm was used.

References

1. Aghdam, M. H., Ghasem-Aghaee, N., & Basiri, M. E. (2009). Text feature selection using ant colony optimization. *Expert Systems with Application, Elsevier, 36*, 6843–6853.
2. Bullnheimer, B., Hartl, R. F., & Strauss, C. (1996). A new rank-based version of the ant system: A computational study. *Central European Journal for Operations Research and Economics, 7*(1), 25–38.
3. Cordon, O., Herrera, F., & Stutzle, T. (2002). A review on the ant colony optimization meta-heuristic: Basis, models and new trends. *Mathware and Soft Computing, 9*(2–3), 141–175.
4. Dash, M., & Liu, H. (1997). Feature selection for classification. *Intelligent Data Analysis, 1*(3), 131–156.
5. Deneubourg, J. L., Goss, S. A. S., & Pasteels, J. M. (1990). The self-organizing exploratory pattern of the argentine ant. *Journal of Insect Behavior, 3*, 159–168.
6. Dorigo, M., & Caro, G. D. (1999). The ant colony optimization metaheuristic. In M. Dorigo & F. Glover (Eds.), *New ideas in optimization* (pp. 11–32). London, UK: McGraw-Hill.
7. Dorigo, M., & Gambardella, L. M. (1997). Ant colony system: A cooperating learning approach to the travelling salesman problem. *IEEE Transactions on Evolutionary Computation, 1*(1), 53–66.
8. Dorigo, M., & Stutzle, T. (2004). *Ant colony optimization*. Cambridge: MIT Press.
9. Forman, G. (2005). Feature selection: We've barely scratched the surface. *IEEE Intelligent Systems, 20*(6), 74–76.
10. Hidalgo, J. M. G. (2003). Text representation for automatic text categorization. In *Tutorial in 10th Conference of the European Chapter of the Association for Computational Linguistics*.
11. Janson, S., Merkle, D., & Middendorf, M. (2005). Parallel ant colony algorithms. In *Parallel Metaheuristics, A New Class of Algorithms*. Chichester: Wiley.
12. Lewis, D. D. (1992). Feature selection and feature extraction for text categorization. In *In the Proceedings of the workshop on Human Language Technology Conference* (pp. 212–217).

13. Li, Y., Luo, C., & Chung, S. M. (2008). Text clustering with feature selection by using statistical data. *IEEE Transactions on Knowledge and Data Engineering, 20*(5), 641–652.
14. Liu, H., & Yu, L. (2005). Towards integrating feature selection algorithms for classification and clustering. *IEEE Transactions on Knowledge and Data Engineering, 17*(4), 491–502.
15. McCallum, A., & Nigam, K. (1998). A comparison of event models for naive bayes text classification. AAAI-98 Workshop on Learning for Text Categorization.
16. Meena, M. J., Chandran, K. R., Karthik, A., & Samuel, A. V. (2012). An enhanced aco algorithm to select features for text categorization and its parallelization. *Expert Systems with Applications: An International Journal, 39*(5), 5861–5871.
17. Sebastiani, F. (2002). Machine learning in automated text categorization. *ACM Computing Surveys, 34*, 1–47.
18. Stutzle, T., & Hoos, H. H. (2000). Maxmin ant system. *Future Generation Computer Systems, 16*(8), 889–914.
19. White, T. (2009). *Hadoop the definitive guide*. Sebastopol, CA: O' Reilly Media Inc.

Chapter 9
The Role of Grid Technologies: A Next Level Combat with Big Data

Manoj Kumar Mishra and Yashwant Singh Patel

Abstract Grid computing has successfully delivered a service oriented architecture that is ubiquitous, dynamic and scalable to the world of networking. It promises to deliver these services to the world of computations that is about to deal with high volume of scalable information involving heterogeneous data i.e., Big data. The Big data needs to explore the new era of technologies and infrastructures that can provide higher level services for managing high volumes of scalable and diverse data. Therefore, it is a timely and challenging opportunity for the Grid technologies to fulfill its promises. This chapter enlightens and examines the key challenges, issues and applications of Grid technologies in the management of Big data.

Keywords Big data · Grid computing · Big data management · Framework · Tools · Technologies

9.1 Introduction

In the last few years the volume of data, i.e. in terms of size, diversified in nature and scalable with respect time has increased radically in various disciplines such as business, manufacturing, science, medical and engineering, etc. Eminent researchers define such a massive heterogeneous data as Big data [19]. These data exhibits high volume and formed with great velocity cannot be structured into the regular database tables [9]. Chen et al. [7] defined Big data as the data sets, whose increasing size is beyond the ability of frequently used software tools to incarcerate, manage and process within a tolerable elapsed time.

Based on the definitions given by researchers, it is observed that there are three main aspects of Big data namely; scale, form and motion. It is estimated that all

M.K. Mishra (✉)
School of Computer Engineering, KIIT University, Bhubaneswar, Odisha, India
e-mail: manojku.mishra05@gmail.com

Y.S. Patel
Department of Computer Science, CDGI, Indore, India
e-mail: yashwant.singh@cdgi.edu.in

© Springer International Publishing Switzerland 2016
B.S.P. Mishra et al. (eds.), *Techniques and Environments for Big Data Analysis*,
Studies in Big Data 17, DOI 10.1007/978-3-319-27520-8_9

the global data spawned from the beginning of time until 2003 corresponds to 5 Exabytes and until 2012, it is nearly 2.7 Zettabytes [9]. Besides, there are several complex and huge computations that cannot be handled even with the help of the supercomputers. Such problems can only be resolved by interconnecting vast amount of heterogeneous resources. The widely used and availability of high-speed networks have dramatically revolutionized the way in which the computation, storage and access to these resources are handled. It is a matter of fact that several million devices like desktops, mobiles, laptops, data vaults, etc. dispersed across the world is handled by billions of users. If all of these devices can be coupled to form a single gigantic and super-powerful computer to manage the demon of Big data, then it is known to be the concept of Grid Computing. As, the environment of Grid computing is inherently dynamic, geographically scattered and heterogeneous in nature, it is the most suitable platform to handle Big data.

9.2   Concerns for Big Data Management

In these days, the amount of data being generated in this world is exploding. The measurement unit of this information is in the range of petabytes or zetabytes. A Hierarchical Representation of Data Scale from Megabytes to Yottabytes [11] is shown in Fig. 9.1. Over the next decade, the storing capacity needed for data will increase by ten times of its current size. Moreover, the practice methodology used to effectively acquire, analyze and interpret such type of unstructured, too big and too fast data will increase the complexity by twice. With the traditional techniques, it would be very difficult to analyze Big data with its varying properties like volume, velocity, variety, variability and value. Since Big data is named to be the most up-to-date upcoming technology in the marketplace that can fetch huge profit to the industry. Hence, it becomes convincible to discover the issues and challenges that are associated with this technology.

Fig. 9.1 A hierarchical representation of data scale from megabyte to yottabyte (log scale) (as reproduced from Ref. [11])

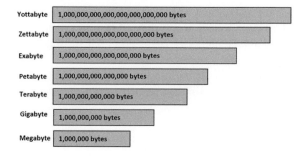

Fig. 9.2 The areas covered by Big data reference architecture (as reproduced from Ref. [16])

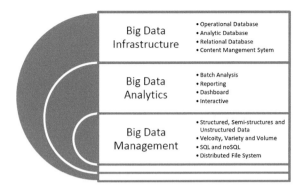

Big Data Infrastructure	• Operational Database • Analytic Database • Relational Database • Content Mangement Sytem
Big Data Analytics	• Batch Analysis • Reporting • Dashboard • Interactive
Big Data Management	• Structured, Semi-structures and Unstructured Data • Velcoity, Variety and Volume • SQL and noSQL • Distributed File System

9.2.1   Big Data

Big data can be defined as the data that is beyond the nominal capability of the traditional database systems and it is a name specified to a new generation of technologies that deals with structured as well as unstructured data for extracting the information by enabling the high-volume, high-velocity and high-variety features [5]. A Big data reference architecture based on its applications is given in Fig. 9.2 [16].

9.2.2   Examples of Big Data

1. RFID (Radio Frequency ID) systems generate up to 1,000 times than the data generated by conventional bar code systems.
2. In each and every second, 10,000 payments are made around the world through card transactions.
3. Walmart handles more than one million customer transactions per an hour.
4. Nearly 4,000 tweets are generated per second, i.e. almost 340 million tweets are dispatched per day.
5. More than 901 million active users of Facebook are generating various types of data through social interaction.
6. Various types of data such as phone calls, messaging, tweeting and glancing websites through android phones are generated by more than 5 billion users [3].

9.2.3   Challenges

The actual concern is not about acquiring huge amount of diverse data; instead it's about what kind of operations are to be performed with the Big data. At present, the main concern is about how to analyze those data and how to retrieve the most valuable pieces of information. The businesses organizations hesitate to make unsophisticated analyses because the absolute volumes of data may devastate their processing platforms. The major concerns that need to be focused while operating with Big data are:

1. How to deal with the varying and large data volumes?
2. Is it needed to store all the data?
3. Is it required to analyze all the data?
4. How to find out the significant data points?
5. How to deal with privacy?
6. How to share the information?

9.2.4   Five V's of Big Data

Big data can be classified and characterized by a multi V model. The multi V model exhibits Variety to represent the types of data, Velocity for processing speed and production of data, Volume to measure the size of data, Veracity to measure the reliability and trust of data and Value to derive the significance while making use of Big data. These five V properties [8, 12] are shown in Fig. 9.3 and are further explained as follows:

(1) V-Variety
All this data is utterly divergent in nature by consisting of raw, structured, semi structured and even unstructured data from different sources like social media sites,

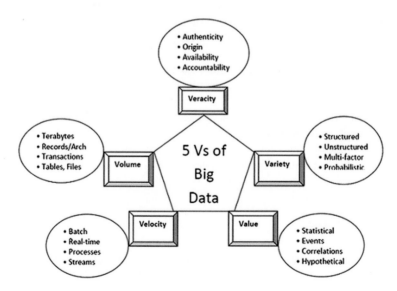

Fig. 9.3 The 5 V's of Big data (as reproduced from Ref. [8, 12])

web pages, and documents etc., which is hard to be handled by the old traditional methodologies.

(2) V-Volume

The term "Big" in the Big data itself describes the volume. Currently, the amount of data generated from various sources is measured in petabytes, but in the near future, it will be measured in zettabytes. The social networking sites produces data in terabytes everyday which represents the high volume of data.

(3) V-Velocity

Velocity represents the notion that deals with data incoming speed from heterogeneous resources. For instance, the data retrieved from the sensor devices is dynamic and this quantity is not tiny enough while storing in database.

(4) V-Variability

Variability refers to many different data types in terms of data formats and data sizes, etc. Such data needs to be maintained properly because of their variability in nature.

(5) V-Value

Users can be able to perform firm queries against the saved data. Thus, it must be able to filter the required data by analyzing the stored data and ordered data should be provided to the users based on their need.

9.2.5   Tools and Technologies Used for Big Data

The most widely used solutions and tools that are used to handle Big data are [20]:

(1) Hadoop: It is the most widely used solution to handle Big data. It was founded by the Apache Software Foundation. It is an open source project based on Google's Map-Reduce and Google File System [2, 10].

(2) Mapreduce: It is the combination of Map and Reduce, where Map() is the master node, which takes the input and divides the input into smaller subparts and distributes it into different worker nodes. On each worker node a recursive operation is performed that leads to the multi-level tree structure. The worker node processes these smaller problems and again returns back the solution to the master node. In Reduce() operation the master node bring together the solutions from all the smaller sub problems and merged them together to form the output.

(3) HBase: It is an open source, Non-relational, and distributed database system written in Java language. It can supply the input as well as the output to MapReduce.

(4) Pig: It is a high-level platform, where the MapReduce programs are created and is applied with Hadoop.

(5) Hive: It is known to be a Data warehousing application that provides the SQL as an interface and a relational model.

(6) Sqoop: It is a command-line interface that is used for transferring of data between relational databases and Hadoop.

(7) Avro: It is generally used in Apache Hadoop for data exchange services.

(8) Oozie: It is a java based web-application that runs in a java servlet. It manages the jobs of Hadoop.

(9) Chukwa: It is a framework for data collection, analysis and processing of large amount logs.

(10) Flume: It is known as a high level architecture that is focused on streaming of data, received from multiple sources.

(11) Zookeeper: It is a centralized service that provides distributed synchronization for group services.

(12) Storm: It is widely used to perform the real time computations because of its scalable, easy to use and fault-tolerant nature [23].

(13) S4: It provides a distributed and scalable platform that is used to process continuous and abundant streams of data [18].

(14) SQLstream: It is a SQL based real-time platform that is used in the field of wireless sensors, M2M communications and for telematics applications [22].

(15) Splunk: Because of its adaptive nature it is widely used in dynamic environment. It is generally used to collect and attach the data of machine [21].

(16) Apache Kafka: It provides a high-throughput stream for fixed motion of data and used for distributed messaging system [1].

(17) SAP Hana: It provides a fast computing platform and widely used for real time businesses [13].

9.2.6 Application Areas of Big Data

As specified by the McKinsey Global Institute, The five broad areas that deals with Big data are [4, 16]:

(i) Public sector:

- To discover the needs
- Customize the actions to enhance the products and services
- Decision assembling with automated systems to decline the risk period.

(ii) Healthcare:

- To provide a clinical decision support system
- Analysis of patient profiling
- Analyze the pattern of diseases

(iii) Individual location of data:

- Smart routing
- Emergency services
- Urban planning

(iv) Manufacturing:

- For supply chain management

- To support sales
- To enhance the applications for web searching

(v) Retailing:

- For store behaviour analysis
- To optimize the variety and price
- Distribution and logistics optimization

9.3   Grid Computing Framework for Big Data Environment

Grid computing can be constructed to be a centralized managed computing environment that is flexible and feasible for high performance computing and faster processing. The major advantage with Grid computing is the large compute intensive capacity, high storage capability and able to manage the workload while maximizing the overall performance. The resources required for the purpose of Big data management should be cohesive in nature.

According to Bote-Lorenzo [6], The Grid characteristics may be described as follows:

(i) Large scale: the Grid deals with a number of heterogeneous resources ranging from just a few to millions.

(ii) Geographical distribution: the resources of Grid may be scattered over distant locations.

(iii) Heterogeneity: the Grid resources both in terms of software and hardware are heterogeneous in nature.

(iv) Resource sharing: the Grid resources may belong to many different adminstrative domains and organizations that allow resource access by other remote organizations. Resource sharing promotes efficiency and reduces costs.

(v) Multiple administrations: the resources of each participating organizations may have different security and access policies under which their owned resources can be used by others.

(vi) Resource coordination: the resources in a Grid computing environment must be coordinated in a manner to provide collective computing capabilities.

(vii) Transparent access: a Grid computing system should be seen as a single virtual computer system.

(viii) Dependable access: the dependable service is fundamental requirement since the users expect predictable, sustained and desired levels of performance. So, the Grid users expect the same Quality of Service (QoS) assurances.

(ix) Consistent access: as Grid is a system, it must have the standarized services, protocols and interfaces for allowing pervasive access and scalability by hiding most of its complex underlaying heterogeneity from the users.

(x) Pervasive access: the resource access should be adaptable to the dynamic environment of the Grid in which appearance of new resources, disappearance of old resources and failure resources is a commonplace.

It is essential to note that Grid uses are not defined in terms of applications they run rather various types of supports the Grid platform provides. From the Grid architecture point of view, the Grid offers following categories of computing services. The major uses of Grid computing are [6, 17]:

- Distributed supercomputing support
- High-throughput computing support
- On-demand computing support
- Data-intensive computing support
- Collaborative computing support
- Multimedia computing support

The environment of Grid computing is inherently dynamic, scalable, geographically scattered and heterogeneous in nature, etc. And by observing the major uses of Grid platform, it is important to notice that Grid computing platform is the most suitable platform to handle Big data. The Grid technologies are successfully implemented for providing the results to the Big data community. In contrast to what is often throught, the Grid platform is not only a computing paradigm, instead, it is an infrastructure that links and unifies the remote and diverse resources dispersed

Fig. 9.4 Grid computing framework for Big data environment

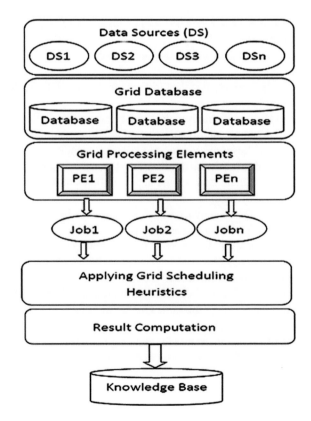

Table 9.1 The latest grid technologies used for Big data

Name of technology	Applied area	Origin	Remarks
Alchemi	Science and technology	Melbourne University, Australia	http://www.cloudbus.org/alchemi/
Condor	Workload management	University of Wisconsin, Madison	http://research.cs.wisc.edu/htcondor/
ClimatePrediction.net	Climate research	Oxford University, England	http://climateprediction.net
LHC@Home	Nuclear research	Europe	http://lhcathomeclassic.cern.ch/sixtrack
Cosmology@Home	Gastronomy	University of Illinois at urban a campaign	http://www.cosmologyathome.org/
Compute against cancer	Cancer research	National Cancer Institute, West Virginia University, University of Maryland	http://globalgridexchange.com/grid/projects/cal.aspx or www.gridcafe.org/
FightAIDS@Home	HIV and AIDS research	Olson laboratory, The Script Research Institute	http://fightaidsathome.org/
Docking@Home	HIV research	National Science foundation, INNOBASE, University of Delaware	http://docking.cis.udel.edu/
HealthGrid	Medical and biomedical applications	France	http://www.healthgrid.org/
TeraGrid	Research and scientific, health science, academic	Sun, IBM, Oracle, HP, Intel	www.xsede.org/home/tg–archives/
NEES	Earthquake researches	United States	https://nees.org/
Bioinformatics Information Research Network (BIRN)	Biomedical science	United States	www.birncommunity.org/

all over the world in order to provide the computing support for a broad range of applications. The generalized Grid computing framework for Big data Environment [14, 15] is depicted in Fig. 9.4.

9.4 Latest Grid Technologies Used for Big Data

The various Grid technologies are summarized in Table 9.1 with their applied areas and origin.

9.5 Conclusion

The world is evolving through a big era of growing size and varieties of data, i.e. the Big data, which requires the support of innovative technologies to deal with it. In this article a brief discussion is presented on Grid computing as a technology to handle Big data, which includes an overview, challenges, applications, and technologies. It is very clear that the current initiatives in this direction, i.e. the techniques and tools applicable to deal with the problems of Big data are awfully limited. Therefore, it is high time for the Government and Enterprises to come up with the investments and further encourage the research on Grid technologies required to handle the challenges of Big data. Big data stands for Big investments, Big Ideas, Big challenges and Big profits. As a technology, the Grid computing has huge potential to manage Big data with lower cost.

References

1. Auradkar, A., et al. (2012). Data infrastructure at LinkedIn. In *IEEE 28th International Conference on Data Engineering (ICDE)*, pp. 1370–1381.
2. Bakshi, K. (2012). Considerations for big data: Architecture and approach. In *IEEE Aerospace Conference*, pp. 1–7.
3. Big data—SAS. http://www.sas.com/en_us/insights/big-data/what-is-big-data.html.
4. Big data technical working groups, white paper. http://big-project.eu/sites/default/files/BIG_D2_2_2.pdf.
5. Bloomberg, J. (2013). Idc definition (conservative and strict approach) of big data, gartner, from the big data long tail blog post by jason bloomberg (2013, Jan 17). http://www.devx.com/blog/the-big-datalong-tail.html, ed dumbill, program chair for the o'reilly strata conference. http://www.gartner.com/it--glossary/big--data/.
6. Bote-Lorenzo, M., Dimitriadis, Y., & Gomez-Sanchez, E. (2004). Grid characteristics and uses: A grid definition (postproceedings extended and revised version). In *Proceedings of the First European Across Grids Conference, ACG'03, Springer, LNCS*, vol. 2970, pp. 291–298.
7. Chen, J., Chen, Y., Du, X., Li, C., Lu, J., Zhao, S., et al. (2013). Big data challenge: A data management perspective. *Frontiers of Computer Science*, 7(2), 157–164.

8. Demchenko, Y. (2012, September). Defining the big data architecture framework (BDAF). *Outcome of the Brainstorming Session at the University of Amsterdam, SNE Group, University of Amsterdam.*

9. Garlasu, D., Sandulescu, V., Halcu, I., Neculoiu, G., Grigoriu, O., Marinescu, M., & Marinescu, V. (2013). A big data implementation based on grid computing. In *The Proceedings of the 11th International Conference on Roedunet*, pp. 1–4.

10. Hadoop. http://searchcloudcomputing.techtarget.com/definition/Hadoop.

11. Harris, R. (2012, September). *International council for science (ICSU) and the challenges of big data in science.* Department of Geography, University College London, Grants, Funding and Science Policy.

12. Katal, A., Wazid, M., & Goudar, R. H. (2013). Big data: Issues, challenges, tools and good practices. In *Sixth International Conference on Contemporary Computing (IC3).*

13. Kraft, S., Casale, G., Jula, A., Kilpatrick, P., & Greer, D. (2012). Wiq: Work-intensive query scheduling for in-memory database systems. In *2012 IEEE 5th International Conference on Cloud Computing (CLOUD)*, pp. 33–40.

14. Ku-Mahamud, K. R. (2013). Big data clustering using grid computing and ant based algorithm. In *4th International Conference on Computing and Informatics, ICOCI*, pp. 6–14.

15. Madheswari, A. N., & Banu, R. S. D. W. (2011). Communication aware co-scheduling for parallel job scheduling in cluster computing. *Advances in Computing and Communications, Communications in Computer and Information Science, 191*, 545–554.

16. Mishra, S. Survey of big data architecture and framework from the industry. In *NIST Big data Public Working Group.*

17. Mishra, M. K., Patel, Y. S., Rout, Y., & Mund, G. (2014). A survey on scheduling heuristics in grid computing environment. *I.J. Modern Education and Computer Science, 10*, 57–83.

18. Neumeyer, L., Robbins, B., Nair, A., & Kesari, A. (2010). S4: Distributed stream computing platform. In *IEEE Data Mining Workshops (ICDMW), Sydney, Australia*, pp. 170–177.

19. Qin, X. (2012). Making use of the big data: Next generation of algorithm trading. In J. Lei, F. Wang, H. Deng & D. Miao (Eds.), *Artificial Intelligence and Computational Intelligence SE 5, 7530*, pp. 34–41, Berlin, Germany: Springer.

20. Sagiroglu, S., & Sinanc, D. (2013). Big data: A review. *International Conference on Collaboration Technologies and Systems (CTS)*, pp. 42–47.

21. Samson, T. (2012). Splunk storm brings log management to the cloud, http://www.infoworld.com/t/managed--services/splunk--storm--brings--logmanagement--the--cloud-201098.

22. Sqlstream. (2012). http://www.sqlstream.com/products/server/.

23. Storm. (2012). http://storm-project.net/.

Printed in the United States
By Bookmasters